Springer Theses

Recognizing Outstanding Ph.D. Research

Aims and Scope

The series "Springer Theses" brings together a selection of the very best Ph.D. theses from around the world and across the physical sciences. Nominated and endorsed by two recognized specialists, each published volume has been selected for its scientific excellence and the high impact of its contents for the pertinent field of research. For greater accessibility to non-specialists, the published versions include an extended introduction, as well as a foreword by the student's supervisor explaining the special relevance of the work for the field. As a whole, the series will provide a valuable resource both for newcomers to the research fields described, and for other scientists seeking detailed background information on special questions. Finally, it provides an accredited documentation of the valuable contributions made by today's younger generation of scientists.

Theses are accepted into the series by invited nomination only and must fulfill all of the following criteria

- They must be written in good English.
- The topic should fall within the confines of Chemistry, Physics, Earth Sciences, Engineering and related interdisciplinary fields such as Materials, Nanoscience, Chemical Engineering, Complex Systems and Biophysics.
- The work reported in the thesis must represent a significant scientific advance.
- If the thesis includes previously published material, permission to reproduce this must be gained from the respective copyright holder.
- They must have been examined and passed during the 12 months prior to nomination.
- Each thesis should include a foreword by the supervisor outlining the significance of its content.
- The theses should have a clearly defined structure including an introduction accessible to scientists not expert in that particular field.

More information about this series at http://www.springer.com/series/8790

Nicolas Maximilian Köhler

Searches for the Supersymmetric Partner of the Top Quark, Dark Matter and Dark Energy at the ATLAS Experiment

Doctoral Thesis accepted by
the Technical University of Munich, Munich,
Germany

 Springer

Author
Dr. Nicolas Maximilian Köhler
Department of Experimental Physics
CERN
Meyrin, Switzerland

Supervisor
Priv.-Doz. Dr. Oliver Kortner
ATLAS group
Max-Planck-Institute for Physics
Munich, Germany

ISSN 2190-5053   ISSN 2190-5061   (electronic)
Springer Theses
ISBN 978-3-030-25990-7   ISBN 978-3-030-25988-4   (eBook)
https://doi.org/10.1007/978-3-030-25988-4

This Springer imprint is published by the registered company Springer Nature Switzerland AG
The registered company address is: Gewerbestrasse 11, 6330 Cham, Switzerland

Supervisor's Foreword

The precision measurements of the anisotropies of the cosmic microwave background radiation deliver clear evidence for the presence of cold dark baryonic matter and dark energy in the universe. None of these are described by the standard model of the strong and electroweak interactions. The lightest supersymmetric particle in R-parity conserving supersymmetric extensions of the standard model provides a natural dark matter candidate. At the same time, the supersymmetric partner of the top quark can solve the standard model's hierarchy problem. Both the production of top squark and generic dark matter models at the LHC lead to the same final state configuration. Nicolas Köhler's thesis exploits this fact and contains limits on the top squark production and the dark matter production cross section. Dark energy in the universe could be the consequence of a scalar field. The final state investigated with ATLAS run-II data in Nicolas Köhler's thesis is sensitive to the production of the corresponding scalar particle. It was therefore possible to set a limit on the existence of this particle which became the world's first limit on dark energy from a collider experiment. Nicolas Köhler's thesis sticks out by the comprehensive interpretation of a particular final state addressing the key questions of contemporary particle and astroparticle physics.

Munich, Germany   Priv.-Doz. Dr. Oliver Kortner
May 2019

Abstract

Astrophysical observations suggest the existence of physics beyond the standard model (SM) of particle physics to explain the composition and evolution of the universe. Super-symmetry (SUSY) is one of the most favoured theoretical frameworks answering the majority of the shortcomings of the SM. In particular, the postulation of supersymmetric particles with differing spin nicely suppresses the unnatural radiative corrections of the Higgs boson mass, and the lightest supersymmetric particle is an ideal dark matter candidate. The former, commonly referred to as the hierarchy problem, predicts the supersymmetric partner of the SM top quark, the so-called top squark, in the TeV mass range. This mass allows for its direct production at the Large Hadron Collider (LHC) at CERN.

This thesis describes the search for direct production of top squark pairs decaying into signatures with jets and missing transverse energy with the ATLAS detector at the LHC. No hint for the existence of top squarks, charginos and neutralinos was found in 36.1 fb^{-1} of proton-proton collision data recorded by the ATLAS experiment in 2015 and 2016 at a centre-of-mass energy of $\sqrt{s} = 13$ TeV. Top squark masses up to 1 TeV are excluded at 95% confidence level (CL) depending on the neutralino mass. Sensitivity studies for the full LHC Run 2 dataset which are performed exploiting Boosted Decision Trees and Artificial Neural Networks show that the analysis presented in this thesis has reached almost the maximum achievable sensitivity.

The same experimental signature as in the search for the top squark is also predicted by more generic extensions of the SM explaining dark matter. Some of these introduce additional spin-0 mediator particles, allowing for the production of weakly interacting massive particles (WIMPs). Mediator masses below 50 GeV can be excluded at 95% CL, assuming a WIMP mass of 1 GeV and a uniform coupling of the mediator to SM particles and WIMPs. For scalar mediators, the exclusion limits are translated into limits on the spin-independent dark matter-nucleon scattering cross section which are compared to the results of direct-detection experiments.

The origin of the accelerated expansion of the universe is one of the most intriguing questions of modern cosmology. The exclusion limits on direct top squark production can be reinterpreted in the context of the production of a scalar particle predicted by a modification of general relativity to explain the origin of dark energy. The results obtained in this thesis are the first collider-based constraints on couplings of dark energy to SM matter.

Muons are key to some of the most important physics results published by the ATLAS experiment including the discovery of the Higgs boson and the measurement of its properties. In a separated part of this thesis, the efficiency of the muon reconstruction and identification is estimated and the performance of high-precision Monitored Drift Tube muon chambers under background rates similar to the ones expected for the High Luminosity-LHC is studied based on $4.0 \, \text{fb}^{-1}$ of proton-proton collision data recorded by the ATLAS experiment in 2016 and 2017 at a centre-of-mass energy of $\sqrt{s} = 13$ TeV.

Acknowledgements

I am deeply grateful to Oliver Kortner for supervising me as his first Ph.D. student. Over the last three years, he supported me and my work in every conceivable way while always keeping in mind the big picture when pointing me to the next stage of my work. Without him, it wouldn't have been possible to obtain the variety of results contained in this thesis in such an efficient way. Furthermore, I want to thank Oliver and also Hubert Kroha, for giving me the opportunity to work at the institute with all its benefits and the great atmosphere within the group. I really enjoyed the last couple of years and I always had the impression to be supported in all my concerns.

I would like to thank Claudia Giuliani for supervising me in the first year of my Ph.D. and teaching me all the secrets of fully-hadronically decaying stops, and I would like to thank Walter Hopkins and Francesca Ungaro for admitting me in their analysis teams in which I had a great time.

I am very grateful to Maximilian Goblirsch-Kolb who brought me to the ATLAS muon community and taught me everything about muon efficiencies and so much other things that I cannot even remember all of them. He still is an exemplary colleague and I enjoy every time working with him. With respect to the muons, I also want to thank Gabriella Sciolla, Massimiliano Bellomo, Stefano Rosati, Federico Sforza and Jochen Meyer who always supported me when I needed their help.

The biggest thanks goes to my office mate Johannes Junggeburth with whom I had an amazing time over the last two years. It was great to spend hours with him developing (more or less needed) code, discussing our next goals, supervising students or even pupils and post-docs and ensuring the successful operation of the beverage supply. I am very grateful that he was thoroughly proofreading my whole thesis and always gave very useful advice.

At the same time, I want to thank my second office mate Rainer Röhrig for the last years, in particular for the legendary times at Ringberg Castle or at the DPGs. But also every working day with you two always was a special pleasure.

I also would like to thank Marvin Lüben, my favourite theoretician when it comes to dark energy and all its mysteries. He was a great support when writing up my thesis. I will really miss the remarkably amazing atmosphere within our group. I thank Catriona Bruce for proofreading my thesis, Korbinian Schmidt-Sommerfeld for teaching me about MDTs, Katharina Ecker for taking care of Rainer and me on trips when we were new to the group, in particular at ESHEP, Verena Walbrecht for her support at our business trips, Dominik Krauss for our soccer discussions in the coffee breaks, Philipp Gadow for his raspberry pie skills and Jonas Graw for being the best Master student I had.

To come back to my life outside physics, I first have to thank my father, Udo, for supporting me from the beginning, making my life as a student so easy and comfortable. I also thank the whole rest of my family for the support over the last years. A very special thanks goes to my friends Sebastian Jäger, Gerrit Illenberger, Jan Jäger, Jens Schauz and Daniel Schmid. I think you are the greatest balance in my life as a physicist and I enjoy every second spent with you. Hopefully, now, I will have more time for you again.

Finally, I have to thank you, Vroni, for being so patient over the last months with me being busy writing all the time. You were the best support, always taking care of me and encouraging me to finalise the thesis. Thank you for that!

Contents

1   Introduction   1

Part I   The Theory of Elementary Particle Physics

2   **The Standard Model of Particle Physics**   5
  2.1   The Electroweak Interaction   7
  2.2   The Strong Interaction   8
  2.3   Electroweak Symmetry Breaking   9
  2.4   Experimental Tests of the Standard Model   11
  2.5   Limitations of the Standard Model   11
  References   13

3   **Dark Matter and Dark Energy**   17
  3.1   The Standard Model of Cosmology   17
  3.2   Observational Evidence for Dark Matter   20
  3.3   Dark Matter Candidates   21
        3.3.1   Weakly Interacting Massive Particles   22
  3.4   Dark Energy and the Expansion of the Universe   22
  References   23

4   **Supersymmetry**   27
  4.1   The Minimal Supersymmetric Standard Model   27
  4.2   R-Parity   29
  4.3   The Hierarchy Problem   29
  4.4   Supersymmetric Dark Matter   31
  4.5   Supersymmetry and the Open Questions of Particle
        Physics   31
  References   32

Part II   The Experimental Setup

5   The Large Hadron Collider at CERN   37
    5.1   Proton–Proton Collisions at the LHC   40
    5.2   The Worldwide LHC Computing Grid   42
    References   43

6   The ATLAS Detector   45
    6.1   The Inner Detector   47
    6.2   The Electromagnetic Calorimeter   48
    6.3   The Hadron Calorimeter   48
    6.4   The Muon Spectrometer   49
    6.5   The Trigger and Data Acquisition System   50
    6.6   Luminosity Measurement   51
    6.7   Pile-Up Interactions   52
    6.8   Physics Object Identification and Reconstruction   53
    References   55

Part III   Performance of Muon Reconstruction and Identification

7   Muon Reconstruction and Identification   61
    7.1   Muon Reconstruction   62
    7.2   Muon Identification   64
    References   66

8   Muon Efficiency Measurement   67
    8.1   Tag-and-Probe Methodology   67
    8.2   Reconstruction Efficiency   68
    8.3   Track-to-Vertex Association Efficiency   77
    References   78

**9   Performance of Monitored Drift Tube Muon Chambers
    at High Rates**   81
    9.1   Estimation of Background Hit Rates   82
    9.2   Spatial Resolution Depending on Background Hit Rates   84
    9.3   Chamber Efficiency Depending on Background Hit Rates   87
    References   89

**Part IV   Searches for New Particles Decaying into Jets and Missing
             Transverse Energy**

10   Searches for New Particles   93
    10.1   Simulation of Standard Model Processes   94
    10.2   General Search Strategy   99
    References   100

11   The Search for the Light Top Squark...................... 103
  11.1   Simulation of Top Squark Pair Production 104
  11.2   Basic Experimental Signature 105
  11.3   Signal Region Definitions 112
  11.4   Background Estimation 130
  11.5   Systematic Uncertainties 145
  11.6   Statistical Interpretation............................. 152
  References 157

12   The Search for Dark Matter.............................. 161
  12.1   Simulation of WIMP Pair Production..................... 162
  12.2   Basic Experimental Signature 163
  12.3   Signal Region Definitions 165
  12.4   Background Estimation 170
  12.5   Systematic Uncertainties 175
  12.6   Statistical Interpretation............................. 175
  References 178

13   The Search for Dark Energy 181
  13.1   Simulation of Dark Energy Production................... 182
  13.2   Choice of Selection Criteria........................... 183
  13.3   Systematic Uncertainties 187
  13.4   Statistical Interpretation............................. 187
  13.5   Validity of the EFT Approximation 187
  References 189

14   Sensitivity Studies Exploiting Multivariate Techniques 191
  14.1   Boosted Decision Trees............................... 192
  14.2   Artificial Neural Networks............................ 200
  14.3   Conclusion 204
  References 205

15   Summary 207

Appendix A: Muon Efficiency Measurements...................... 209

Appendix B: Search for Direct Top Squark Pair Production 221

Appendix C: Search for Dark Matter 245

Appendix D: Search for Dark Energy............................. 251

**Appendix E: Multivariate Techniques in the Search
              for Top Squarks** 253

Chapter 1
Introduction

Modern physics tries to solve the mysteries of our universe such as its origin, composition and future evolution. Physicists are searching for fundamental laws to consistently describe all existing particles and their interactions. The Standard Model of particle physics is one of the most successful theoretical models and was intensively probed by a plethora of precision measurements. Despite all experimental measurements being in excellent agreement with the theoretical predictions and the last missing piece of the Standard Model, the Higgs boson, being discovered in 2012, there is no doubt that the Standard Model is insufficient to fully describe astroparticle physics. Aside from the fact that the Standard Model does not describe the gravitational interaction, there is astrophysical evidence for the existence of Dark Matter and Dark Energy, which make up 95% of the universe but cannot be associated to any of the Standard Model particles. Furthermore, from a theoretical point of view, the Higgs boson at 125 GeV is unnaturally light compared to the energy scale where we expect gravity to be quantised.

Supersymmetry postulates the existence of superpartners differing in spin-1/2 for every Standard Model particle, which could explain the measured Higgs boson mass in case the mass of the superpartner of the top quark, the top squark, is in the TeV range and therefore accessible by the Large Hadron Collider (LHC) at CERN. Furthermore, the lightest supersymmetric particle is stable and serves as an ideal Dark Matter candidate. This thesis presents the search for the top squark, with subsequent decays into the neutralino and additional jets, based on $36.1\,\mathrm{fb}^{-1}$ of proton-proton collision data taken at the ATLAS experiment at a centre-of-mass energy of $\sqrt{s} = 13\,\mathrm{TeV}$.

However, in case Supersymmetry is not realised in nature, one should search for Dark Matter in a more generic way. Currently, the most favoured Dark Matter candidate is the Weakly Interacting Massive Particle (WIMP). Light WIMPs can be directly produced at the LHC via the exchange of a spin-0 mediator and result in experimental signatures with large missing transverse momenta. Theoretical models assuming a coupling of the mediator to the Standard Model Higgs potential suggest that the coupling strength is proportional to the Yukawa couplings and thus the

© Springer Nature Switzerland AG 2019
N. M. Köhler, *Searches for the Supersymmetric Partner of the Top Quark,
Dark Matter and Dark Energy at the ATLAS Experiment*,
Springer Theses, https://doi.org/10.1007/978-3-030-25988-4_1

production of the mediator in association with top quark pairs is preferred, which results in a similar signature to top squark production.

The accelerated expansion of the universe is commonly associated with the presence of Dark Energy, which makes up 69% of universe's energy content. Its origin is one of the most intriguing subjects of modern cosmology. A subset of theoretical models explaining the existence of Dark Energy with modified gravity introducing a scalar field coupling to Standard Model particles claims that Dark Energy can be produced at the LHC in collisions with large momentum transfer involving top quark pair production. In this thesis, for the first time, accelerator-based constraints on Dark Energy models are obtained by reinterpreting the results from the search for top squarks.

The work focuses on experimental signatures with jets and large missing transverse momenta, which benefit from higher branching ratios than decays involving leptons.

The basic theoretical concepts of the Standard Model of particle physics, Dark Matter and Dark Energy are discussed in the first part of the thesis. The LHC and the ATLAS detector are then introduced.

In a separated part of the thesis, the performance of the muon reconstruction and identification of the ATLAS detector is estimated, which is a key ingredient for precision measurements with muon signatures. Furthermore, the performance of high-precision Monitored Drift Tube muon chambers depending on the background hit rates present in the ATLAS cavern is studied with respect to the detector upgrades associated with the High Luminosity-LHC.

The fourth and main part of the thesis presents the ATLAS searches for the top squark as well as Dark Matter and Dark Energy including a dedicated chapter on the application of multivariate techniques such as Boosted Decision Trees and Artificial Neural Networks in high-energy physics.

Part I
The Theory of Elementary Particle Physics

Chapter 2
The Standard Model of Particle Physics

It is remarkable that the vast number of physics observations made up to this day can be explained by four fundamental interactions: The *gravitational interaction* making large masses attract each other, the *electromagnetic interaction* acting on electrically charged particles, the *weak interaction* responsible for radioactive decays and the *strong interaction* keeping together the nuclei. The inherent randomness of quantum mechanics paired with Einstein's Theory of Special Relativity [1] has led to the introduction of *Quantum Field Theory* as the mathematical language of particle physics.

This chapter describes the combination of the electromagnetic, weak and strong interactions in the *Standard Model of particle physics*. Surprisingly, it only describes approximately 5% of the content of our universe [2] while the rest is designated Dark Matter or Dark Energy. The gravitational force [3] which dominates at cosmic scales becomes negligible in particle physics experiments on Earth since it is—being around 26 orders of magnitude weaker than the weak force [4]—by far the weakest of all fundamental interactions.

The *Standard Model of cosmology* predicts the existence of Dark Matter and Dark Energy [5]. It is introduced in Chap. 3 focussing on the most popular Dark Matter candidate, the *Weakly Interacting Massive Particle*. The phenomenology of scalar Dark Energy signatures at colliders, needed to understand the results shown in Chap. 13, will also be discussed.

Supersymmetry, which is one of the most promising extensions of the Standard Model of particle physics, will be introduced in Chap. 4. It is capable of explaining almost all of the shortcomings of the Standard Model of particle physics in an elegant way and serves as the key prerequisite of string theory—the approach to unify all forces in a *grand unified theory* [6].

The Standard Model (SM) of particle physics [7–9] is a relativistic quantum field theory of the interactions of elementary fermions via the electroweak and the strong force based on the local gauge symmetry group SU(3) $\times$ SU(2) $\times$ U(1) [10]. In the second half of the 20th century, the presence of different neutrino flavours as well as the observation of symmetries with the charged leptons hinted at the

© Springer Nature Switzerland AG 2019
N. M. Köhler, *Searches for the Supersymmetric Partner of the Top Quark, Dark Matter and Dark Energy at the ATLAS Experiment*, Springer Theses, https://doi.org/10.1007/978-3-030-25988-4_2

existence of an extension of Quantum electrodynamics (QED) including the weak force (cf. Sect. 2.1). Around the same time, the discovery of various new unstable hadrons [11–13] led to the development of *Quantum chromodynamics*, a gauge theory describing the strong interaction between colour-charged particles (cf. Sect. 2.2). The mechanism of electroweak symmetry breaking which is explained in Sect. 2.3 was introduced to preserve the gauge invariance of the electroweak Lagrangian in the presence of massive weak gauge bosons and massive fermions. The SM was strictly tested and all experimental observations show remarkable agreement as described in Sect. 2.4. The remaining shortcomings of the Standard Model of particle physics will be discussed in Sect. 2.5.

Quantum field theories rely on the Lagrangian formalism of classical field theory. There, the Lagrangian density $\mathcal{L}(\varphi)$ allows for the derivation of the equations of motion of a physical system as the Euler-Lagrange equations from Hamilton's principle [14]

$$\partial_\mu \frac{\partial \mathcal{L}}{\partial \left(\partial_\mu \varphi \right)} - \frac{\partial \mathcal{L}}{\partial \varphi} = 0 \ . \tag{2.1}$$

Elementary particles are described by fields that are operators on the quantum mechanical Fock space which allows to describe many-body systems in quantum mechanics [15]. In the Fock state basis, several indistinguishable particles can occupy on single-particle Fock state such that exchanging two particles does not lead to a different many-body state. In order to add or remove a particle from the many-body system, so-called *creation* and *annihilation operators* are defined. The field operators are linear superpositions of the creation and annihilation operators.

The Particle Content of the Standard Model

All particles can be classified into two classes according to their spin: Particles with integer spin are *bosons* while particles with half-integer spin are *fermions*. The Standard model contains spin-0 and spin-1 bosons and spin-$\frac{1}{2}$ fermions. The fermion fields ψ are the constituents of matter and are mathematically described by spinors. They can be divided into leptons and quarks, each occurring in 3 generations.

Gauge Invariance in the Standard Model

In order to construct the gauge invariant quantum field theory of the Standard Model, the local gauge transformation of a fermion field,

$$\psi \rightarrow \exp\left(i\alpha^a(x^\mu)T_a \right) \psi \tag{2.2}$$

with the local phase $\alpha^a(x^\mu)$ has to be accompanied by transforming the gauge field as [14]

$$A_\mu^a \rightarrow A_\mu^a + \frac{1}{g} \cdot \partial_\mu \alpha^a (x^\mu) + f_{abc} A_\mu^b \alpha^c (x^\mu) \quad . \tag{2.3}$$

T_a ($a \in \{1, \ldots, N\}$) are the generators of the corresponding symmetry group $SU(n)$ where $N = n^2 - 1$, g is the coupling strength of the interaction and f_{abc} ($a, b, c \in \{1, \ldots, N\}$) are the so-called group structure constants which fulfil the commutation relation

$$[T_a, T_b] = i f_{abc} T_c \quad . \tag{2.4}$$

The gauge invariance is ensured by replacing the partial derivative ∂_μ by the covariant derivative

$$\partial_\mu \rightarrow D_\mu = \partial_\mu - \frac{1}{2} i g T_a A_\mu^a \tag{2.5}$$

in the Lagrangian of the theory.

2.1 The Electroweak Interaction

The electromagnetic and weak interactions were unified by Glashow, Salam and Weinberg [7–9] resulting in the electroweak theory using the symmetry group

$$SU(2)_L \times U(1)_Y \quad , \tag{2.6}$$

where $SU(2)_L$ ($U(1)_Y$) is the symmetry group of the weak (electromagnetic) interaction and Y is the weak hypercharge. The L indicates that the $SU(2)_L$ gauge fields only couple to left-handed fermions. Right-handed fermions are $SU(2)_L$-singlets and do not interact weakly within the SM.[1] The weak hypercharge Y is a conserved quantum number in the SM and is related to the electric charge Q and the third component of the weak isospin I_3 by the Gell–Mann–Nishijima formula [16]

$$Q = I_3 + \frac{Y_W}{2} \quad . \tag{2.7}$$

The three generators of the $SU(2)_L$ gauge group result in the existence of three gauge boson fields $W_\mu^{1,2,3}$. One vector boson field B_μ is associated to the $U(1)_Y$ group. According to Eq. (2.5), the covariant derivative of the electroweak interaction can be written as

$$D_\mu = \partial_\mu - \frac{1}{2} i g_W \sigma_a W_\mu^a - \frac{1}{2} i g_Y Y B_\mu^a \quad , \tag{2.8}$$

---

[1]Since neutrinos only interact weakly, right-handed neutrinos do not interact within the SM at all.

where g_W (g_Y) is the weak (electromagnetic) coupling constant and σ_a ($a \in \{1, 2, 3\}$) are the Pauli matrices, the generators of $SU(2)_L$. The Lagrangian of the electroweak interaction

$$\mathcal{L}_{EW} = \sum_f \bar{f} i \gamma^\mu D_\mu f - \frac{1}{4} F^a_{\mu\nu} F^{\mu\nu}_a \tag{2.9}$$

contains the interaction term of fermions f with the electroweak gauge bosons and the kinetic term of the electroweak gauge fields, with $F^a_{\mu\nu}$ being the electroweak field strength tensor

$$F^a_{\mu\nu} = \partial_\mu W^a_\nu - \partial_\nu W^a_\mu + g_W f_{abc} W^b_\mu W^c_\nu + \partial_\mu B^a_\nu - \partial_\nu B^a_\mu \quad . \tag{2.10}$$

Since adding mass terms for the electroweak gauge fields to Eq. (2.10) would break the gauge invariance, the experimental observation of massive $W^\pm$ and Z bosons requires the introduction of electroweak symmetry breaking (cf. Sect. 2.3).

2.2   The Strong Interaction

The strong interaction is described by Quantum chromodynamics (QCD) based on the $SU(3)_C$ symmetry group where C denotes the colour-charge. The only fermions carrying colour charge are quarks which are triplets under $SU(3)_C$. The colours in QCD are commonly labelled red, green and blue—referring to the RBG-colour scheme. Leptons are singlets under $SU(3)_C$ and thus, do not carry any colour-charge. According to Eq. (2.5), the covariant derivative of QCD can be written as

$$D_\mu = \partial_\mu - \frac{1}{2} i g_s \lambda_a G^a_\mu \quad , \tag{2.11}$$

where g_s is the strong coupling constant—often referred to as α_s—and λ_a ($a \in \{1, \ldots, 8\}$) are the Gell-Mann matrices, the generators of $SU(3)_C$. Hence, QCD has eight gauge bosons G^a_μ which are called gluons.

Analogously to the electroweak interaction, the kinetic terms of the strong gauge fields are obtained by

$$\mathcal{L}_{QCD,\text{kin.}} = -\frac{1}{4} G^a_{\mu\nu} G^{\mu\nu}_a \quad , \tag{2.12}$$

with $F^a_{\mu\nu}$ being the strong strength tensor

$$G^a_{\mu\nu} = \partial_\mu G\nu^a - \partial_\nu G^a_\mu + g_s f_{abc} G^b_\mu G^c_\nu \quad . \tag{2.13}$$

Equation (2.12) leads to the interaction of gluon fields, i.e. triple and quartic gluon self-interactions become possible in QCD.

2.3 Electroweak Symmetry Breaking

As mentioned in Sect. 2.1, the gauge invariance requires the electroweak gauge bosons to be massless. To explain the experimental evidence of massive $W^{\pm}$ and Z bosons, an additional complex scalar field

$$\phi = \begin{pmatrix} \phi^+ \\ \phi^0 \end{pmatrix} = \frac{1}{\sqrt{2}} \begin{pmatrix} \phi_3 + i\phi_4 \\ \phi_1 + i\phi_2 \end{pmatrix} \quad , \tag{2.14}$$

is introduced [17, 18] which transforms as a $SU(2)_L$ doublet and is called the *Higgs field*. It carries four degrees of freedom and has the electroweak quantum numbers $I_3 = \pm\frac{1}{2}$ and $Y = 1$. The Higgs field is introduced into the SM Lagrangian [19] by adding

$$\mathcal{L}_{\text{Higgs}} = \left(D_\mu \phi\right)^\dagger \left(D^\mu \phi\right) - \left(\mu^2 \phi^\dagger \phi + \lambda \left(\phi^\dagger \phi\right)^2\right) \tag{2.15}$$

where D_μ is the covariant derivative of the electroweak interaction (Eq. (2.8)) and the second term is called the *Higgs potential* $V(\phi)$. In case of $\lambda > 0$ and $\mu^2 < 0$, the Higgs potential has its non-trivial minima at

$$\min\{V(|\phi|)\} = \sqrt{-\frac{\mu^2}{2\lambda}} \tag{2.16}$$

which corresponds to a vacuum expectation value of

$$\langle 0| \phi |0\rangle = \sqrt{-\frac{\mu^2}{2\lambda}} \equiv \frac{v}{\sqrt{2}} \neq 0 \quad . \tag{2.17}$$

Since the neutral photon is the only massless electroweak boson, the explicit minimum of the Higgs potential is chosen such that the charged component of the Higgs doublet is zero while the neutral one acquires the vacuum expectation value,

$$\phi_0 = \frac{1}{\sqrt{2}} \begin{pmatrix} 0 \\ v \end{pmatrix} \quad . \tag{2.18}$$

The choice of one specific realisation breaks the $SU(2)_L \times U(1)_Y$ symmetry leaving the $U(1)_Y$ symmetry of the electromagnetic interaction unbroken. This mechanism is called *spontaneous electroweak symmetry breaking*.

Expanding the Higgs field around the chosen minimum in unitarity gauge leads to

$$\phi = \frac{1}{\sqrt{2}} \begin{pmatrix} 0 \\ v + H \end{pmatrix} \tag{2.19}$$

where H is a massive excitation which is called the Higgs boson and has a mass of $m_H = \sqrt{-\mu^2}$. Inserting Eq. (2.19) into the kinetic term of Eq. (2.15) and diagonal-ising the mass matrix results in the photon and Z boson fields

$$\begin{pmatrix} Z_\mu \\ A_\mu \end{pmatrix} := \begin{pmatrix} \cos(\theta_W) & \sin(\theta_W) \\ -\sin(\theta_W) & \cos(\theta_W) \end{pmatrix} \cdot \begin{pmatrix} W_\mu^3 \\ B_\mu \end{pmatrix} \tag{2.20}$$

where $\cos(\theta_W)$ is defined by [8]

$$\cos(\theta_W) := \frac{g_W}{\sqrt{g_W^2 + g_Y^2}} \tag{2.21}$$

with θ_W being the Weinberg angle. Defining $W_\mu^\pm := \frac{1}{\sqrt{2}}(W_\mu^1 \mp i W_\mu^2)$, the masses of the electroweak gauge bosons can be written as

$$m_W = \frac{g_W \cdot v}{2} \quad \text{and} \quad m_Z = \frac{g_W \cdot v}{2\cos(\theta_W)} \quad . \tag{2.22}$$

The presence of the Higgs fields ϕ finally allows to write down the mass terms of the quarks and the charged leptons. Extending the SM Lagrangian by

$$\mathcal{L}_{\text{Yukawa}} = -y_\ell^{ij} \ell_{Li}^\dagger e_{Rj} \phi - y_q^{ij} q_{Li}^\dagger u_{Rj} i\sigma_2 \phi^* - y_q^{ij} q_{Li}^\dagger d_{Rj} \phi + h.c. \tag{2.23}$$

where y^{ij} are the Yukawa coupling matrices, σ_2 is the second Pauli matrix and $i \in \{1, 2, 3\}$ are the three fermion generations, keeps the Lagrangian invariant under $SU(2)_L \times U(1)_Y$. Note that Eq. (2.23) does not contain mass terms for neutrinos, since right handed neutrinos are not part of the SM. Thus, within the SM, neutrinos have to be massless.

Since the Yukawa coupling matrices are non-diagonal, the mass eigenstates of fermions are mixtures of the electroweak eigenstates. For quarks, the unitary Cabibbo–Kobayashi–Maskawa (CKM) matrix [20, 21] serves as the mass mixing matrix.

All fermions of the Standard Model of particle physics are summarised in Table 2.1. The colour-charge of the quarks is not represented, since it would triple the number of quarks shown in the Table.

All in all, the SM with the approximation of massless neutrinos can be fully described by 19 free parameters: 6 quark masses, 3 lepton masses, 3 gauge coupling constants (g_Y, g_W, g_s), the Higgs boson mass m_H and self-coupling strength λ, 4 parameters of the CKM-matrix and one CP-violating parameter for the strong interaction [22].

Table 2.1 Elementary fermions of the SM of particle physics with their electroweak quantum numbers

	Fermions			I	I_3	Q	Y
Leptons	$\begin{pmatrix} \nu_e \\ e \end{pmatrix}_L$	$\begin{pmatrix} \nu_\mu \\ \mu \end{pmatrix}_L$	$\begin{pmatrix} \nu_\tau \\ \tau \end{pmatrix}_L$	$\frac{1}{2}$	$+\frac{1}{2}$ $-\frac{1}{2}$	0 -1	-1
	e_R	μ_R	τ_R	0	0	-1	-2
Quarks	$\begin{pmatrix} u \\ d \end{pmatrix}_L$	$\begin{pmatrix} c \\ s \end{pmatrix}_L$	$\begin{pmatrix} t \\ b \end{pmatrix}_L$	$\frac{1}{2}$	$+\frac{1}{2}$ $-\frac{1}{2}$	$+\frac{2}{3}$ $-\frac{1}{3}$	$\frac{1}{3}$
	u_R	c_R	t_R	0	0	$+\frac{2}{3}$	$+\frac{4}{3}$
	d_R	s_R	b_R	0	0	$-\frac{1}{3}$	$-\frac{2}{3}$

2.4 Experimental Tests of the Standard Model

The vast majority of particle physics experiments so far confirmed the SM predictions with high precision [23, 24]. The electroweak gauge bosons were discovered at CERN[2] in 1983 [25]. The heaviest quark, the top quark, was discovered at the Tevatron in 1994 [26]. All its properties agree with the SM prediction [27, 28]. The last missing piece of the SM, the Higgs boson, was discovered by the ATLAS and CMS experiments at the Large Hadron Collider at CERN (cf. Chap. 5) in 2012 [29, 30]. Also the measurements of the properties of the Higgs boson show no deviations from the SM yet [31–45]. Another remarkable test of the SM is depicted in Fig. 2.1 which shows the world-averages of the measured $W^\pm$ boson and top quark mass (green) as well as the simultaneous indirect determination of $m_{W^\pm}$ and m_t using precision tests of the electroweak and strong interactions. The blue area is including the measured Higgs boson mass in the global electroweak fit, while the grey area is excluding it. Both contours agree with the direct measurements within the uncertainties which demonstrates the consistency of the SM. Furthermore, the fit indirectly determines the $W^\pm$ boson mass [46] to

$$m_{W^\pm,\text{theo.}} = 80.358 \pm 0.008 \, \text{GeV}. \tag{2.24}$$

This precisely agrees with the $W^\pm$ boson mass recently measured in ATLAS [47]

$$m_{W^\pm,\text{meas.}} = 80.370 \pm 0.019 \, \text{GeV}. \tag{2.25}$$

2.5 Limitations of the Standard Model

Despite the overwhelming agreement with the measurements, the SM suffers from theoretical shortcomings. Besides the fact that the SM does not include a quantum

---

[2]The European Organisation for Nuclear Research in Geneva, Switzerland, originally: *Conseil Européen pour la Recherche Nucléaire.*

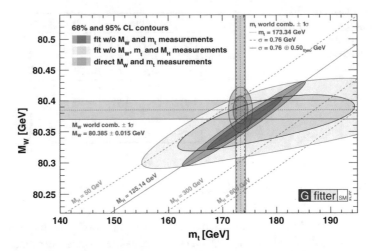

Fig. 2.1 Contours of 68% and 95% confidence level obtained from scans of fits with fixed variable pairs $m_{W^{\pm}}$ versus m_t. The narrower blue and larger grey allowed regions are the results of the fit including and excluding the m_H measurements, respectively. The horizontal bands indicate the 1σ regions of the $m_{W^{\pm}}$ and m_t measurements (world averages) [46]

field theoretic description of gravity, the main motivations for physics beyond the SM are:

- The observations of neutrino oscillations [48] proves the existence of non-vanishing neutrino masses.
- Astrophysical observations [49–52] indicate the existence of matter which does not interact electromagnetically and therefore is called Dark Matter [53] as discussed in Chap. 3.
- The observation that the expansion of our universe is accelerating [2] can only be explained by introducing the so-called Dark Energy [54] whose particle nature is completely unknown (cf. Chap. 3).
- The asymmetry between matter and antimatter formed in the early universe cannot be explained by the observed CP violation in the weak interaction [55, 56].
- Assuming the existence of physics beyond the SM, the rather small value of 125 GeV for the Higgs boson mass [32, 57] causes another problem: The energy scale of electroweak symmetry breaking is of the order of 100 GeV which is much smaller than the $\sim 10^{18}$ GeV of the Planck scale [58]. Since the scalar Higgs boson couples to every massive particle, its mass receives radiative corrections Δm_H^2 from quantum loop processes which are proportional to the chosen cut-off energy scale of an effective field theory [59]. Since the observed Higgs mass

$$m_H^2 = \left(m_{H,\text{bare}}\right)^2 + \Delta m_H^2 \tag{2.26}$$

consists of the bare mass $m_{H,\text{bare}}$ and its radiative corrections, it would require an unnatural *fine-tuning* of the bare mass and the corrections over plenty orders of magnitude. This theoretically motivated problem is referred to as the *hierarchy problem* which will be further discussed in Sect. 4.3.

- Although the strong interaction also allows for CP violation, it is experimentally constrained to be very small by measurements of the neutron dipole moment [60]. This aesthetical problem of QCD is also referred to as the strong CP problem [61, 62].
- The fact that above a certain energy, the electromagnetic and weak force unify in the electroweak interaction described by common coupling constants g_Y and g_W (cf. Sect. 2.1) paired with the discovery that the coupling constant of the strong interaction α_s is decreasing with increasing energy scales [63, 64] suggest that QCD could also unify with the electroweak interaction at higher energies. Theoretical models describing this unification are called *Grand Unified Theories* and require extending of the SM [65].

References

1. Einstein A (2005) Zur Elektrodynamik bewegter Körper [AdP 17, 891 (1905)]. Annalen der Physik 14(S1):194–224. https://doi.org/10.1002/andp.200590006
2. Ade PAR et al (2014) Planck 2013 results. I. Overview of products and scientific results. Astron Astrophys 571:A1. https://doi.org/10.1051/0004-6361/201321529
3. Einstein A (1916) The foundations of the theory of general relativity. German. AdP 354(7):769–822. https://doi.org/10.1002/andp.19163540702
4. Greiner W, Müller B (2009) Gauge theory of weak interactions, 4th edn. Springer, Berlin, p 2
5. Dodelson S (2003) Modern cosmology. Academic Press, Elsevier Science, Cambridge
6. Green MB, Schwarz JH, Witten E (1987) Superstring theory. Cambridge University Press, Cambridge
7. Glashow SL (1961) Partial symmetries of weak interactions. Nucl Phys 22:579–588. https://doi.org/10.1016/0029-5582(61)90469-2
8. Steven W (1967) A model of leptons. Phys Rev Lett 19:1264–1266. https://doi.org/10.1103/PhysRevLett.19.1264
9. Salam A (1968) Elementary particle theory. Almqvist and Wiksell, Sweden
10. Srednicki M (2007) Quantum field theory. Cambridge University Press, Cambridge
11. Rochester GD, Butler CC (1947) Evidence for the existence of newunstable elementary particles. Nature 160:855–857. https://doi.org/10.1038/160855a0
12. Powell CF, Occhialini GPS (1947) Nuclear physics in photographs: tracks of charged particles in photographic emulsions. Clarendon Press, Oxford. http://cds.cern.ch/record/1030134
13. Chamberlain O et al (1955) Observation of antiprotons. Phys Rev 100:947–950. https://doi.org/10.1103/PhysRev.100.947
14. Hollik W (2010) Quantum field theory and the Standard Model. In: High-energy physics. Proceedings of 17th European School, ESHEP 2009, Bautzen, Germany, June 14–27, 2009. arXiv:1012.3883[hep-ph]
15. Fock V (1932) Konfigurationsraum und zweite Quantelung. Zeitschrift für Physik 75(9):622–647. https://doi.org/10.1007/BF01344458
16. Nakano T, Nishijima K (1953) Charge independence for V-particles*. Prog Theor Phys 10(5):581–582. https://doi.org/10.1143/PTP.10.581
17. Englert F, Brout R (1964) Broken symmetry and the mass of gauge vector mesons. Phys Rev Lett 13:321–323. https://doi.org/10.1103/PhysRevLett.13.321

18. Higgs PW (1964) Broken symmetries and the masses of gauge bosons. Phys Rev Lett 13:508–509. https://doi.org/10.1103/PhysRevLett.13.508
19. Higgs PW (1966) Spontaneous symmetry breakdown without massless bosons. Phys Rev 145:1156–1163. https://doi.org/10.1103/PhysRev.145.1156
20. Cabibbo N (1963) Unitary symmetry and leptonic decays. Phys Rev Lett 10:531–533. https://doi.org/10.1103/PhysRevLett.10.531
21. Kobayashi M, Maskawa T (1973) CP violation in the renormalizable theory of weak interaction. Prog Theor Phys 49:652–657. https://doi.org/10.1143/PTP.49.652
22. Ellis J (2000) Standard model of particle physics. Encycl Astron Astrophys. Institute of Physics Publishing. https://doi.org/10.1888/0333750888/2104
23. Schael S et al (2013) Electroweak measurements in electron-positron collisions at w-boson-pair energies at LEP. Phys Rep 532:119–244. https://doi.org/10.1016/j.physrep.2013.07.004
24. Schael S et al (2006) Precision electroweak measurements on the Z resonance. Phys Rep 427:257–454. https://doi.org/10.1016/j.physrep.2005.12.006
25. Arnison G et al (1983) Experimental observation of lepton pairs of invariant mass around 95 GeV/c2 at the CERN SPS collider. Phys Lett B 126:398–410. https://doi.org/10.1016/0370-2693(83)90188-0
26. Abe F et al (1995) Observation of top quark production in $\bar{p}p$ collisions. Phys Rev Lett 74:2626–2631. https://doi.org/10.1103/PhysRevLett.74.2626
27. ATLAS Collaboration (2014) Measurement of the top-quark mass in $t = \bar{t}$ events with lepton+jets final states in pp collisions at $\sqrt{s} = 8TeV$. Technical report CMS-PAS-TOP-14-001. Geneva: CERN
28. ATLAS, CDF, CMS and DØ Collaborations (2014) First combination of Tevatron and LHC measurements of the top-quark mass. ATLAS-CONF-2014-008
29. ATLAS Collaboration (2012) Observation of a new particle in the search for the standard model Higgs boson with the ATLAS detector at the LHC. Phys Lett B 716:1. https://doi.org/10.1016/j.physletb.2012.08.020
30. CMS Collaboration (2012) Observation of a new boson at a mass of 125 GeV with the CMS experiment at the LHC. Phys Lett B 716:30. https://doi.org/10.1016/j.physletb.2012.08.021
31. ATLAS Collaboration (2013) Measurements of Higgs boson production and couplings in diboson final states with the ATLAS detector at the LHC. Phys Lett B 726:88. https://doi.org/10.1016/j.physletb.2014.05.011
32. ATLAS Collaboration (2014) Measurement of the Higgs boson mass from the $H \to \gamma\gamma$ and $H \to ZZ^*4\ell$ channels in pp collisions at center-of-massenergies of 7 and 8 TeV with the ATLAS detector. Phys Rev D 90:052004. https://doi.org/10.1103/PhysRevD.90.052004
33. ATLAS Collaboration (2012) Coupling properties of the new Higgs-like boson observed with the ATLAS detector at the LHC. ATLAS-CONF-2012-127. https://cds.cern.ch/record/1476765
34. ATLAS Collaboration (2012) Updated results and measurements of properties of the new Higgs-like particle in the four lepton decay channel with the ATLAS detector. ATLAS-CONF-2012-169. https://cds.cern.ch/record/1499628
35. ATLAS Collaboration (2013) Measurements of the properties of the Higgs-like boson in the two photon decay channel with the ATLAS detector using 25 fb^1 of proton-proton collision data. ATLAS-CONF-2013-012. https://cds.cern.ch/record/1523698
36. ATLAS Collaboration (2013) Measurements of the properties of the Higgs-like boson in the four lepton decay channel with the ATLAS detector using 25 fb^1 of proton-proton collision data. ATLAS-CONF-2013-013. https://cds.cern.ch/record/1523699
37. ATLAS Collaboration (2013) Measurements of the properties of the Higgs-like boson in the $WW^{(*)} \to \ell\nu\nu$ decay channel with the ATLAS detector using 25 fb^1 of proton-proton collision data. ATLAS-CONF-2013-030. https://cds.cern.ch/record/1527126
38. ATLAS Collaboration (2013) Study of the spin properties of the Higgs-like boson in the $H \to WW(*) \to e\nu\mu\nu$ channel with 21 fb^1 of $\sqrt{s} = 8TeV$ data collected with the ATLAS detector. ATLAS-CONF-2013-031. https://cds.cern.ch/record/1527127
39. ATLAS Collaboration (2016) Study of the Higgs boson properties and search for high-mass scalar resonances in the $H \to ZZ^* \to 4\ell$ decay channel at $\sqrt{s} = 13TeV$ with the ATLAS detector. ATLAS-CONF-2016-079. https://cds.cern.ch/record/2206253

40. ATLAS Collaboration (2017) Measurement of the Higgs boson coupling properties in the $H \rightarrow ZZ^* \rightarrow 4\ell$ decay channel at $\sqrt{s} = 13TeV$ with the ATLAS detector. ATLAS-CONF-2017-043. https://cds.cern.ch/record/2273849
41. ATLAS Collaboration (2017) Measurements of Higgs boson properties in the diphoton decay channel with 36.1 fb^{-1} pp collision data at the center-of-mass energy of $13TeV$ with the ATLAS detector. ATLAS-CONF-2017-045. https://cds.cern.ch/record/2273852
42. CMS Collaboration (2014) Observation of the diphoton decay of the Higgs boson and measurement of its properties. Eur Phys J C 74:3076. https://doi.org/10.1140/epjc/s10052-014-3076-z
43. CMS Collaboration (2014) Measurement of the properties of a Higgs boson in the four-lepton final state. Phys Rev D 89:092007. https://doi.org/10.1103/PhysRevD.89.092007
44. CMS Collaboration (2014) Measurement of Higgs boson production and properties in the WW decay channel with leptonic final states. JHEP 01:096. https://doi.org/10.1007/JHEP01(2014)096
45. CMS Collaboration (2017) Measurements of properties of the Higgs boson decaying into the four-lepton final state in pp collisions at $\sqrt{s} = 13TeV$. JHEP 11:047. https://doi.org/10.1007/JHEP11(2017)047
46. Baak M, Kogler R (2013) The global electroweak standard model fit after the Higgs discovery. In: Proceedings of 48th Rencontres de Moriond on Electroweak Interactions and Unified Theories: La Thuile, Italy, March 2–9, pp 349–358. arXiv: 1306.0571[hep-ph]
47. ATLAS Collaboration (2018) Measurement of the W-boson mass in pp collisions at $\sqrt{s} = 7TeV$ with the ATLAS detector. Eur Phys J C 78:110. https://doi.org/10.1140/epjc/s10052-017-5475-4
48. Forero DV, Tortola M, Valle JWF (2012) Global status of neutrino oscillation parameters after Neutrino-2012. Phys Rev D 86:073012. https://doi.org/10.1103/PhysRevD.86.073012
49. Zwicky F (1933) Spectral displacement of extra galactic nebulae. Helv Phys Acta 6:110–127
50. Begeman KG, Broeils AH, Sanders RH (1991) Extended rotation curves of spiral galaxies: dark haloes and modified dynamics. Mon Not R Astron Soc 249:523
51. Komatsu E et al (2009) Five-year wilkinson microwave anisotropy probe observations: cosmological interpretation. Astrophys J 180(2):330
52. Bertone G, Hooper D, Silk J (2005) Particle dark matter: evidence, candidates and constraints. Phys Rep 405:279–390. https://doi.org/10.1016/j.physrep.2004.08.031
53. Feng JL (2010) Dark matter candidates from particle physics and methods of detection. Annu Rev Astron Astrophys 48(1):495–545. https://doi.org/10.1146/annurev-astro-082708-101659
54. Frieman J, Turner M, Huterer D (2008) Dark energy and the accelerating universe. Ann Rev Astron Astrophys 46:385–432. https://doi.org/10.1146/annurev.astro.46.060407.145243
55. Sakharov AD (1967) Violation of CP invariance, asymmetry, and baryon asymmetry of the universe. Pisma Zh Eksp Teor Fiz 5:32–35. https://doi.org/10.1070/PU1991v034n05ABEH002497
56. Cline JM (2000) Status of electroweak phase transition and baryogenesis. Pramana 55:33–42. https://doi.org/10.1007/s12043-000-0081-6
57. CMS Collaboration (2015) Precise determination of the mass of the Higgs boson and tests of compatibility of its couplings with the standard model predictions using proton collisions at 7 and 8 TeV. Eur Phys J C 75:212. https://doi.org/10.1140/epjc/s10052-015-3351-7
58. Burdman G (2007) New solutions to the hierarchy problem. Braz J Phys 37:506–513. https://doi.org/10.1590/S0103-97332007000400006
59. Hooft G et al (1980) Recent developments in gauge theories. In: Nato advanced study institute, Cargese, France, August 26–September 8, 1979, vol. 59
60. Harris PG et al (1999) New experimental limit on the electric dipole moment of the neutron. Phys Rev Lett 82:904–907. https://doi.org/10.1103/PhysRevLett.82.904
61. Pospelov M, Ritz A (2005) Electric dipole moments as probes of new physics. Ann Phys 318:119–169. https://doi.org/10.1016/j.aop.2005.04.002
62. Kim JE, Carosi G (2010) Axions and the strong CP problem. Rev Mod Phys 82:557–602. https://doi.org/10.1103/RevModPhys.82.557

63. Gross DJ, Wilczek F (1973) Ultraviolet behavior of non-abelian gauge theories. Phys Rev Lett 30:1343–1346. https://doi.org/10.1103/PhysRevLett.30.1343
64. David Politzer H (1974) Asymptotic freedom: an approach to strong interactions. Phys Rep 14:129–180. https://doi.org/10.1016/0370-1573(74)
65. de Boer W (1994) Grand unified theories and supersymmetry in particle physics and cosmology. Prog Part Nucl Phys 33:201–302. https://doi.org/10.1016/0146-6410(94)90045-0

Chapter 3
Dark Matter and Dark Energy

The existence of Dark Matter would explain several astrophysical observations not covered by the Standard Model of particle physics. An even more intangible and unknown form of energy, which is supposed to explain the accelerating expansion of our universe, Dark Energy, completes the *dark sector* of particle physics which makes up approximately 95% of our universe [1]. This chapter introduces the Standard Model of cosmology in Sect. 3.1 which gives the foundation for a variety of theoretical models for Dark Matter and Dark Energy. In Sect. 3.2, the most important observational evidences for the existence of Dark Matter are introduced, before Sect. 3.3.1 focusses on the model of *weakly interacting massive particles* as Dark Matter candidate, which will be part of the searches presented in Chaps. 11 and 12. Finally, Sect. 3.4 discusses one specific model of Dark Energy which could be potentially produced at earthbound particle colliders like the Large Hadron Collider (LHC) at CERN.

3.1 The Standard Model of Cosmology

Over the course of the past several decades, the increasing number of accurate astrophysical observations led to the establishment of the Standard Model of cosmology. It can be considered as the counterpart of the SM of particle physics, describing the evolution of our universe as a whole by only considering effects on cosmological scales. The Standard Model of cosmology is based on three key assumptions [2]:

- The universe is *homogeneous* and *isotropic* on scales larger than roughly $150\,\text{Mpc} \approx 4.6 \cdot 10^{24}\,\text{m}$.
- The energy content of the universe, such as ordinary matter, Dark Matter and Dark Energy is modelled in terms of cosmological fluids.
- Gravitational interactions between those fluids are described by Einstein's Theory of General Relativity [3].

© Springer Nature Switzerland AG 2019
N. M. Köhler, *Searches for the Supersymmetric Partner of the Top Quark, Dark Matter and Dark Energy at the ATLAS Experiment*, Springer Theses, https://doi.org/10.1007/978-3-030-25988-4_3

The assumption of a homogeneous and isotropic universe is also called the *Cosmological principle* [4]. It is based on the *Copernican principle* assuming that we are not located at an extraordinary position in the universe. Homogeneity cannot be directly confirmed, since astrophysical observations can only be made in the past light cone. The experimental evidence for isotropy observed in the Cosmic Microwave Background [5], X-ray background [6] and radio galaxies [7], in addition to the Copernican principle however implies the homogeneity of the universe [4, 8].

After measurements of the light spectra of nearby galaxies showed that they are moving away [9, 10], Hubble found that their radial velocity v_{rad} is proportional to their distance r from Earth [11]

$$v_{\text{rad}} = H_0 \cdot r \quad . \tag{3.1}$$

The constant of proportionality $H_0 = 67.8 \pm 0.9\,\text{km}\,\text{s}^{-1}\,\text{Mpc}^{-1}$ [12] is called the *Hubble* constant. The cosmological principle infers that Hubble's law (Eq. (3.1)) is valid everywhere and thus, the universe is expanding. Around the same time, Friedmann independently found that the universe might expand by solving Einstein's field equations of general relativity [3]

$$R_{\mu\nu} - \frac{1}{2}g_{\mu\nu}R = -\frac{8\pi G_N}{c^4}T_{\mu\nu} - \Lambda g_{\mu\nu} \quad , \tag{3.2}$$

where $R_{\mu\nu}$ and R are the Ricci tensor and Ricci scalar, respectively, and $g_{\mu\nu}$ is the metric tensor. On the right hand side, G_N is Newton's constant, $T_{\mu\nu}$ is the energy-momentum tensor of all SM fields and Λ the *cosmological constant* artificially introduced by Einstein in order to achieve a static universe. Einstein's field equations relate the geometry of the universe (left hand side of Eq. (3.2)) to its energy content (right hand side). Friedmann derived his equations from the *Friedmann–Lemaître–Robertson–Walker metric* [13–16]

$$g_{\mu\nu}\mathrm{d}x^\mu\mathrm{d}x^\nu = -c^2\mathrm{d}s^2 = -c^2\mathrm{d}t^2 + a(t)^2\left(\frac{\mathrm{d}r^2}{1 - kr^2} + r^2\mathrm{d}\Omega^2\right) \quad , \tag{3.3}$$

which is a special solution to Einstein's field equations in polar coordinates and thus, a unification of General Relativity and the cosmological principle. The constant $k \in \{-1, 0, +1\}$ is determining the *topology* of the universe. For $k = -1$, the metric is an open 3-hyperboloid and the universe would be *open*, for $k = 0$, the metric is of Euclidean space and the universe would be *flat* while for $k = +1$, the metric is a closed 3-sphere and the universe would be *closed*. Inserting the metric from Eq. (3.3) into the field equations (3.2), one obtains the Friedmann equations [13]

$$\left(\frac{\dot{a}(t)}{a(t)}\right)^2 + \frac{kc^2}{a(t)^2} = \frac{8\pi G_N}{3} \cdot \rho_{\text{tot}}(t) \quad, \tag{3.4}$$

$$\frac{\ddot{a}(t)}{a(t)} = -\frac{4\pi G_N}{3}\left(\rho_{\text{tot}}(t) + \frac{3p_{\text{tot}}(t)}{c^2}\right) \quad, \tag{3.5}$$

where $\rho_{\text{tot}}(t)$ and $p_{\text{tot}}(t)$ are the total energy density and pressure of the universe, respectively.

The relation between $\rho_{\text{tot}}(t)$ and $p_{\text{tot}}(t)$ is given via the equation of state [2]

$$p_{\text{tot}}(t) = \omega(\rho)\rho_{\text{tot}}(t) \quad, \tag{3.6}$$

where $\omega(\rho)$ is called the barotropic parameter. The total energy density can be written as

$$\rho_{\text{tot}} = \rho_{\text{matter}} + \rho_{\text{radiation}} + \rho_\Lambda \quad, \tag{3.7}$$

where ρ_{matter} is the matter density, $\rho_{\text{radiation}}$ is the radiation density and ρ_Λ is the so-called *vacuum energy*. By setting $k = 0$ and solving Eq. (3.4) to $\rho_{\text{tot}}(t)$, the critical density

$$\rho_{\text{crit}}(t) \equiv \rho_{\text{tot}}(t)\Big|_{k=0} = \frac{3H(t)^2}{8\pi G_N} \tag{3.8}$$

is defined which is used to define the *density parameter* [2]

$$\Omega_i(t) = \frac{\rho_i(t)}{\rho_{\text{crit}}(t)} \tag{3.9}$$

for a given type of energy fluid $i \in \{\text{matter, radiation, ...}\}$. Hubble's law (Eq. (3.1)) can be derived from the Friedmann equations

$$\frac{\dot{a}(t)}{a(t)} = H(t) \quad, \tag{3.10}$$

making the Hubble constant to a function of time. The value of $H(t)$ at present time t_0 corresponds to the Hubble constant $H(t_0) = H_0$. The density parameters of today's matter fluid $\Omega_{\text{matter}}(t_0)$ is determined from Cosmic Microwave Background measurements [12] to be $\Omega_{\text{matter}}(t_0) = 0.308 \pm 0.012$. However, only $\Omega_{\text{b. matter}}(t_0) = 0.0483 \pm 0.0005$ of that matter density is baryonic matter[1] [12]. Thus, the particle nature of roughly 26% of the universe' matter density is unknown. Since it does not emit electromagnetic radiation, it is usually referred to as Dark Matter.

Since astrophysical observations indicate that the universe is almost flat [17], the total energy density of the universe is expected to be one. Thus, Eq. (3.7) can be

---

[1]On astronomical scales, also accompanying electrons inside atoms are part of baryonic matter.

rearranged and the density parameter of the vacuum energy is given by ($\Omega_{radiation}(t_0)$ is negligible)

$$\Omega_\Lambda(t_0) = 1 - \Omega_{matter}(t_0) - \Omega_{radiation}(t_0) \approx 69\% \quad . \tag{3.11}$$

Indeed, the density parameter of the vacuum energy was measured to be $\Omega_\Lambda(t_0) = 0.685 \pm 0.013$ [12]. The exceptional property of the vacuum energy is that the barotropic parameter of its equation of state (Eq. (3.6)) is negative which causes the accelerated expansion of the universe. This completely unknown form of energy is denoted Dark Energy and will be further discussed in Sect. 3.4.

3.2 Observational Evidence for Dark Matter

First evidence for the existence of Dark Matter came from the observations of the *rotation curves of spiral galaxies*. In Newtonian dynamics, the circular velocity v of stars or surrounding gases as a function of the distance r to the galactic centre is given by

$$v(r) = \sqrt{\frac{G_N M(r)}{r}} \tag{3.12}$$

where $M(r) = \int \rho(r) r^2 dr$ is the mass of the galaxy with mass density $\rho(r)$ inside a virtual sphere with radius r from its centre. Using the distributions of luminous objects, the circular velocity is expected to fall proportional to $1/\sqrt{r}$ at large distances beyond the galactic disc, however it was measured to be constant [18, 19]. In case the validity of Newtonian dynamics is assumed, it implies the existence of an halo with $M(r) \propto r$ and $\rho(r) \propto r^{-2}$ [20]. The optical invisibility of the mass halo leads to the assumption of the existence of Dark Matter.

Einstein's General Theory of Relativity implies that very massive objects in space are bending light from more distant sources and therefore act as lenses for the observer. This effect is called *gravitational lensing*. Here, one distinguishes between *strong* gravitational lensing caused by galaxy clusters which lead to the appearance of more than one image of the original source [21] and *weak* gravitational lensing which only causes distortions of the image from the original source [22]. Both kinds of lensing are only sensitive to the gravitational interaction, thus also the existence of Dark Matter will contribute.

On extragalactic scales, galaxy clusters originally gave the first indications for the existence of Dark Matter. The mass of galaxy clusters can be estimated in three independent ways. Firstly, using weak gravitational lensing; secondly, measuring the X-ray energy spectrum and flux of the cluster and calculating the gas temperature and pressure and assuming that it is balanced with the gravitational collapse; or, applying the virial theorem to the observed radial velocities of the galaxies. Comparing the three estimations of the galaxy cluster mass gives insight into the relation of matter and visible matter. In 1933, Zwicky measured the mass-to-light ratio of the Coma cluster clearly exceeding the ratio of our galaxy [23]. Today, there is a variety of observations consistent with $\Omega_{matter}(t_0)$ measured from the Cosmic Microwave Background [24–27].

While the observation of the rotation curves of spiral galaxies and the weak gravitational lensing can also be explained by a modification of Newton's laws (MOND) [28], the observation of strong gravitational lensing imposes strong constraints on the validity of MOND theories [29]. Furthermore, no satisfactory cosmological model based on MOND theories has been constructed so far [30]. Thus, the existence of Dark Matter in the context of the Standard Model of cosmology is the currently most favoured theoretical model.

3.3 Dark Matter Candidates

There are several candidates for the particle nature of Dark Matter. Since Dark Matter does not emit electromagnetic radiation, the only possible baryonic Dark Matter candidates are black holes, neutron stars or brown dwarfs which are commonly called *massive compact halo objects* (MACHOs) [31]. However, MACHOs cannot explain the full amount of Dark Matter present in the universe [32]. Dark Matter is assumed to be non-baryonic and therefore, it can at most interact via the electroweak and gravitational forces. This also implies that a particle candidate for Dark Matter requires an extension of the SM of particle physics.

The most popular candidates for non-baryonic Dark Matter are:

- Neutrinos: Since neutrinos are massless in the SM of particle physics, an extension would become necessary. However, their total relic density is much too low to explain the amount of Dark Matter in the universe [20].
- Sterile neutrinos: Massive right-handed neutrinos which are singlets under the electroweak gauge group $SU(2)_L$ are Dark Matter candidates [33, 34]. However, since they only interact gravitationally, they are difficult to observe and there is no experimental evidence for the existence of sterile neutrinos yet [35].

Depending on how far Dark Matter particles could move due to random motions in the early universe, one can distinguish between *cold*, *warm* and *hot* Dark Matter. While neutrinos qualify as either of the three types, the following candidates are part of the so-called Λ-*Cold Dark Matter* (ΛCDM) model, where Λ is the cosmological constant.

- Axions: The scalar particles which were postulated to solve the strong CP problem (cf. Sect. 2.5) are possible Dark Matter candidates [36]. They are already experimentally constrained to be very light [37, 38].
- Weakly interacting massive particles (WIMPs): Particles interacting via the gravitational and electroweak forces, or potentially via a weaker force not contained in

the SM of particle physics, are commonly called *weakly interacting massive particles*. Requiring the correct abundance of Dark Matter today, the typical WIMP mass is expected to be roughly in the $\mathcal{O}(100\,\text{GeV})$ mass range [39]. A short introduction to WIMPs will be given in Sect. 3.3.1.

- Light scalar Dark Matter: If scalar Dark Matter is assumed, their mass range can be reduced to 1–100 MeV [40] which could potentially explain the astrophysical observation of the 511 keV gamma-ray line [41].
- There is a variety of other Dark Matter candidates such as Little Higgs models [42–45], Kaluza-Klein states [46], Superheavy dark matter [47, 48] or Q-balls [49, 50] which will not be discussed in this thesis.

3.3.1 Weakly Interacting Massive Particles

Weakly interacting massive particles (WIMPs) are assumed to remain as a relic of the Big Bang. In the dense environment of the early universe, all particles were in thermal equilibrium which means their interaction rates were larger than the expansion of the universe. As the universe expands, its temperature decreases and, after dropping below the Dark Matter particle mass, the interaction of Dark Matter with other SM particles such as photons becomes exponentially suppressed [51]. Thus, the abundance of Dark Matter approaches a constant value called the *thermal relic abundance*. From the large scale structure of the universe, constraints on the Dark Matter abundance, mass and lifetime can be derived [52]. To explain the cosmological abundance of Dark Matter in today's universe, the Dark Matter annihilation cross section has to be of the order 1 pb which is in the range of a typical electroweak cross section and hints to a Dark Matter mass in the range of $\mathcal{O}(100\,\text{GeV})$ [51, 53]. One of the most promising WIMP candidates is coming from the introduction of Supersymmetry [54] which will be discussed in Chap. 4. However, the WIMP does not have to be a supersymmetric particle, WIMPs are also predicted by *universal extra dimensions* (the lightest Kaluza-Klein particle) [46] or *Little Higgs* theories [42–45].

3.4 Dark Energy and the Expansion of the Universe

Cosmological observations reveal that the expansion of the universe is currently accelerating [55, 56]. This effect is commonly attributed to the presence of Dark Energy whose origin is completely unknown. The cosmological constant introduced by Einstein to explain a flat universe is the simplest modification to the General Theory of Relativity that can act as Dark Energy and is in very good agreement with all present experimental observations. However, interpreting Dark Energy as the vacuum energy in the total energy density of the universe (cf. Eq. (3.7)), leads to theoretical shortcomings similar to those of the hierarchy problem discussed in Sect. 4.3. Those can be resolved by a scale dependent modification of the General

Theory of Relativity which becomes visible at large cosmological scales [57]. These so-called modified gravity theories would have to reduce to Einstein's field equations for galactic scales to be in agreement with all current experimental observations.

Theories of modified gravity are commonly introducing new scalar fields whose potential energies may cause the accelerated expansion of the universe [58]. The most general scalar-tensor theories coupling to matter without fields with negative kinetic energy are referred to as *Horndeski theories* [59]. Since no such scalar particle has been observed thus far, the scale dependence of the theory can be realised by either some environment-depending screening mechanism or by imposing a shift symmetry which forbids Yukawa type interactions between SM fermions and the scalar field [60].

Recently, Brax et al. proposed that the realisation of a shift symmetric Horndeski theory coupling to SM particles could be constrained using experimental LHC data [60]. For scalar Dark Energy masses of $m_\varphi = 0.1\,\text{GeV}$, the pair production of Dark Energy scalars φ in association with top quark pair production ($2\varphi + t\bar{t}$) is expected to share a similar phenomenology as the production of Dark Matter (cf. Chap. 13).

References

1. Landim RCG (2017) Dark sector cosmology. PhD thesis, Sao Paulo U. arXiv: 1702.06024 [gr-qc]
2. Pettinari GW (2015) The intrinsic bispectrum of the cosmic microwave background. Springer theses, Springer International Publishing. isbn: 9783319218823
3. Einstein A (1916) The foundations of the theory of general relativity, German. AdP 354(7):769–822. https://doi.org/10.1002/andp.19163540702
4. Maartens R (2011) Is the universe homogeneous? Philos Trans R Soc Lond A369:5115–5137. https://doi.org/10.1098/rsta.2011.0289
5. Bennett CL et al (1996) Four year COBE DMR cosmic microwave background observations: maps and basic results. Astrophys J 464:L1–L4. https://doi.org/10.1086/310075
6. Scharf CA et al (2000) The 2–10 keV XRB dipole and its cosmological implications. Astrophys J 544:49. https://doi.org/10.1086/317174
7. Peebles PJE (1994) Principles of physical cosmology
8. Wartofsky MW, Cohen RS (1984) In: Cohen RS, Wartofsky MW (eds) Physical sciences and history of physics. English: D. Reidel; Distributed in the USA and Canada by Kluwer Academic Publishers Dordrecht; Boston: Hingham, MA, USA, ix, 259 p. isbn: 9027716153
9. Slipher VM (1913) The radial velocity of the Andromeda Nebula. Lowell Obs Bull 2:56–57
10. Slipher VM (1915) Spectrographic observations of nebulae. Popul Astron 23:21–24
11. Hubble E (1929) A relation between distance and radial velocity among extragalactic nebulae. Proc Natl Acad Sci 15(3):168-173. issn: 0027-8424. https://doi.org/10.1073/pnas.15.3.168
12. Ade PAR et al (2016) Planck 2015 results. XIII. Cosmological parameters. Astron Astrophys 594:A13. https://doi.org/10.1051/0004-6361/201525830
13. Friedmann A (1922) Über die Krümmung des Raumes. Z Phys 10:377–386. https://doi.org/10.1007/BF01332580
14. Lemaître G (1931) Expansion of the universe, a homogeneous universe of constant mass and increasing radius accounting for the radial velocity of extra-galactic nebulae. Mon Not R Astron Soc 91:483–490. https://doi.org/10.1093/mnras/91.5.483

15. Robertson HP (1935) Kinematics and world-structure. Astrophys J 82:284. https://doi.org/10.
    1086/143681
16. Walker AG (1937) On Milne's theory of world-structure. Proc Lond Math Soc s2-42(1):90-127.
    https://doi.org/10.1112/plms/s2-42.1.90
17. Hinshaw G et al (2013) Nine-year Wilkinson microwave anisotropy probe (WMAP) obser-
    vations: cosmological parameter results. Astrophys J Suppl 208:19. https://doi.org/10.1088/
    0067-0049/208/2/19
18. Rubin VC, Ford WK Jr (1970) Rotation of the Andromeda Nebula from a spectroscopic survey
    of emission regions. Astrophys J 159:379. https://doi.org/10.1086/150317
19. Rubin VC, Thonnard N, Ford WK Jr (1978) Extended rotation curves of high-luminosity spiral
    galaxies. IV—Systematic dynamical properties, SA through SC. Astrophys J 225:L107–L111.
    https://doi.org/10.1086/182804
20. Bertone G, Hooper D, Silk J (2005) Particle dark matter: evidence, candidates and constraints.
    Phys Rep 405:279–390. https://doi.org/10.1016/j.physrep.2004.08.031
21. Taylor AN et al (1998) Gravitational lens magnification and the mass of Abell 1689. Astrophys
    J 501:539. https://doi.org/10.1086/305827
22. Hoekstra H, Yee H, Gladders M (2002) Current status of weak gravitational lensing. New
    Astron Rev 46:767–781. https://doi.org/10.1016/S1387-6473(02)00245-2
23. Zwicky F (1933) Die Rotverschiebung von extragalaktischen Nebeln. Helv Phys Acta 6:110–
    127
24. Bahcall NA (2000) Cosmology with clusters of galaxies. Phys Scr 2000(T85):32
25. Kashlinsky A (1998) Determining Σ from the cluster correlation function. Phys Rep 307(1):67–
    73. issn: 0370-1573. https://doi.org/10.1016/S0370-1573(98)00050-7
26. Clowe D, Gonzalez A, Markevitch M (2004) Weak lensing mass reconstruction of the in-
    teracting cluster 1E0657-558: direct evidence for the existence of dark matter. Astrophys J
    604:596–603. https://doi.org/10.1086/381970
27. Markevitch M et al (2004) Direct constraints on the dark matter self-interaction cross-section
    from the merging galaxy cluster 1E0657-56. Astrophys J 606:819–824. https://doi.org/10.
    1086/383178
28. Milgrom M (1983) A modification of the Newtonian dynamics as a possible alternative to the
    hidden mass hypothesis. Astrophys J 270:365–370. https://doi.org/10.1086/161130
29. Pointecouteau E, Silk J (2005) New constraints on MOND from galaxy clusters. Mon Not R
    Astron Soc 364:654–658. https://doi.org/10.1111/j.1365-2966.2005.09590.x
30. Scott D et al (2001) Cosmological difficulties with modified Newtonian dynamics (or, La Fin
    du MOND?). Mon Not R Astron Soc (submitted for publication)
31. Bennett DP et al (1996) The MACHO project dark matter search. ASP Conf Ser 88:95
32. Raine D, Thomas EG (2001) An introduction to the science of cosmology. Series in astronomy
    and astrophysics. Taylor & Francis. isbn: 9780750304054
33. Kusenko A, Takahashi F, Yanagida TT (2010) Dark matter from split seesaw. Phys Lett B
    693:144–148. https://doi.org/10.1016/j.physletb.2010.08.031
34. Krauss LM, Nasri S, Trodden M (2003) A model for neutrino masses and dark matter. Phys
    Rev D 67:085002. https://doi.org/10.1103/PhysRevD.67.085002
35. Aartsen MG et al (2016) Searches for sterile neutrinos with the icecube detector. Phys Rev Lett
    117(7):071801. https://doi.org/10.1103/PhysRevLett.117.071801
36. Rosenberg LJ, van Bibber KA (2000) Searches for invisible axions. Phys Rep 325:1–39. https://
    doi.org/10.1016/S0370-1573(99)00045-9
37. Peccei RD (2008) The strong CP problem and axions. Lect Notes Phys 741:3–17. https://doi.
    org/10.1007/978-3-540-73518-2_1
38. Anastassopoulos V et al (2017) New CAST limit on the axion-photon interaction. Nat Phys
    13:584–590. https://doi.org/10.1038/nphys4109
39. Lee BW, Weinberg S (1977) Cosmological lower bound on heavy-neutrino masses. Phys Rev
    Lett 39:165–168. https://doi.org/10.1103/PhysRevLett.39.165
40. Boehm C, Fayet P (2004) Scalar dark matter candidates. Nucl Phys B 683:219–263. https://
    doi.org/10.1016/j.nuclphysb.2004.01.015

41. Chun EJ, Kim HB (2006) Axino light dark matter and neutrino masses with R-parity violation. JHEP 10:082. https://doi.org/10.1088/1126-6708/2006/10/082
42. Arkani-Hamed N, Cohen AG, Georgi H (2001) Electroweak symmetry breaking from dimensional deconstruction. Phys Lett B 513:232–240. https://doi.org/10.1016/S0370-2693(01)00741-9
43. Arkani-Hamed N et al (2002) Phenomenology of electroweak symmetry breaking from theory space. JHEP 08:020. https://doi.org/10.1088/1126-6708/2002/08/020
44. Arkani-Hamed N et al (2002) The minimal moose for a little Higgs. JHEP 08:021. https://doi.org/10.1088/1126-6708/2002/08/021
45. Arkani-Hamed N et al (2002) The littlest Higgs. JHEP 07:034. https://doi.org/10.1088/1126-6708/2002/07/034
46. Agashe K, Servant G (2004) Warped unification, proton stability and dark matter. Phys Rev Lett 93:231805. https://doi.org/10.1103/PhysRevLett.93.231805
47. Kolb EW, Chung DJH, Riotto A (1999) WIMPzillas! AIP Conf Proc 484(1):91–105. https://doi.org/10.1063/1.59655
48. Chang S, Coriano C, Faraggi AE (1996) Stable superstring relics. Nucl Phys B 477:65–104. https://doi.org/10.1016/0550-3213(96)00371-9
49. Kusenko A et al (1998) Experimental signatures of supersymmetric dark matter Q balls. Phys Rev Lett 80:3185–3188. https://doi.org/10.1103/PhysRevLett.80.3185
50. Kusenko A, Shaposhnikov ME (1998) Supersymmetric Q balls as dark matter. Phys Lett B 418:46–54. https://doi.org/10.1016/S0370-2693(97)01375-0
51. Kamionkowski M (1997) WIMP and axion dark matter. In: High-energy physics and cosmology. Proceedings, Summer School, Trieste, Italy, 2 June–4 July 1997, pp 394–411
52. Steigman G, Turner MS (1985) Cosmological constraints on the properties of weakly interacting massive particles. Nucl Phys B 253:375–386. issn: 0550-3213. https://doi.org/10.1016/0550-3213(85)90537-1
53. Salati P (2014) Dark matter annihilation in the universe. Int J Mod Phys Conf Ser 30:1460256. https://doi.org/10.1142/S2010194514602567
54. Jungman G, Kamionkowski M, Griest K (1996) Supersymmetric dark matter. Phys Rep 267(5):195–373. issn: 0370-1573. https://doi.org/10.1016/0370-1573(95)00058-5
55. Perlmutter S et al (1999) Measurements of Omega and Lambda from 42 high redshift supernovae. Astrophys J 517:565–586. https://doi.org/10.1086/307221
56. Riess AG et al (1998) Observational evidence from supernovae for an accelerating universe and a cosmological constant. Astron J 116:1009–1038. https://doi.org/10.1086/300499
57. Brax P (2018) What makes the universe accelerate? A review on what dark energy could be and how to test it. Rep Prog Phys 81(1):016902
58. Joyce A et al (2015) Beyond the cosmological standard model. Phys Rep 568:1–98. https://doi.org/10.1016/j.physrep.2014.12.002
59. Horndeski GW (1974) Second-order scalar-tensor field equations in a four-dimensional space. Int J Theor Phys 10:363–384. https://doi.org/10.1007/BF01807638
60. Brax P et al (2016) LHC signatures of scalar dark energy. Phys Rev D 94(8):084054. https://doi.org/10.1103/PhysRevD.94.084054

Chapter 4
Supersymmetry

Supersymmetry (SUSY) is the currently most favoured theory beyond the SM. The symmetry generators Q of SUSY transform bosonic states into fermionic states and vice versa [1–7],

$$Q\,|\text{boson}\rangle = |\text{fermion}\rangle\,, \qquad Q\,|\text{fermion}\rangle = |\text{boson}\rangle\,, \tag{4.1}$$

and fulfil the anticommutation relations

$$\{Q, Q^\dagger\} = P^\mu, \tag{4.2}$$

$$\{Q, Q\} = \{Q^\dagger, Q^\dagger\} = 0, \tag{4.3}$$

where P^μ is the four-momentum operator generating space-time translations. The introduction of the symmetry generators postulates the existence of a supersymmetric partner for every SM particle which are grouped into supermultiplets.

If SUSY were an exact symmetry, all properties of supersymmetric particles apart from their spin would be the same as their SM partners. This implies that SUSY must be broken since, for instance, no superpartner of the electron has been observed so far. Spontaneous SUSY breaking can take place as in the electroweak interaction resulting in a hidden symmetry at low energies [7].

4.1 The Minimal Supersymmetric Standard Model

The Minimal Supersymmetric Standard Model (MSSM) is the minimal supersymmetric extension of the SM of particle physics [8, 9]. Since the MSSM is based on the same gauge group as the SM, there are 8 colour vector supermultiplets V^α, 3 weak supermultiplets V^i and a hypercharge singlet V which are all shown in Table 4.1.

© Springer Nature Switzerland AG 2019
N. M. Köhler, *Searches for the Supersymmetric Partner of the Top Quark,
Dark Matter and Dark Energy at the ATLAS Experiment*,
Springer Theses, https://doi.org/10.1007/978-3-030-25988-4_4

Table 4.1 The vector supermultiplets of the MSSM [10]. In the third column, the spin-1 gauge bosons as well as their spin-$\frac{1}{2}$ superpartners are shown

Supermultiplet	$SU(3) \times SU(2) \times U(1)$	Particles
V^α	$(8, 1, 0)$	Gluons g_α and gluinos $\tilde{g}_\alpha$ $(\alpha = 1, \ldots, 8)$
V^i	$(1, 3, 0)$	W_i and winos $\tilde{W}_i$ $(i = 1, 2, 3)$
V	$(1, 1, 0)$	B and bino $\tilde{B}$

Table 4.2 The chiral supermultiplets of the MSSM [10] with the fermions and their scalar superpartners, called sfermions, and with the 2 Higgs doublets and the supersymmetric spin-$\frac{1}{2}$ Higgsinos

Supermultiplet	$SU(3) \times SU(2) \times U(1)$	Particles
Q	$(3, 2, +\frac{1}{3})$	Quarks (u, d) and squarks $(\tilde{u}, \tilde{d})$
$\bar{U}$	$(\bar{3}, 1, -\frac{4}{3})$	Quarks $(\bar{u})$ and squarks $(\bar{\tilde{u}})$
$\bar{D}$	$(\bar{3}, 1, +\frac{2}{3})$	Quarks $(\bar{d})$ and squarks $(\bar{\tilde{d}})$
L	$(1, 2, -1)$	Leptons (ν, e) and sleptons $(\tilde{\nu}, \tilde{e})$
$\bar{E}$	$(1, 1, +2)$	Electron (e) and selectron $(\bar{\tilde{e}})$
H_1	$(1, 2, -1)$	Higgs doublet (h_1) and Higgsino doublet $(\tilde{H}_1)$
H_2	$(1, 2, +1)$	Higgs doublet (h_2) and Higgsino doublet $(\tilde{H}_2)$

The supermultiplets corresponding to the quarks and leptons are shown in Table 4.2. The Higgs sector contains 2 Higgs doublets[1] which lead to 8 degrees of freedom compared to 4 in the SM. After electroweak symmetry breaking, 3 of them give the masses to $W^\pm$ and Z while 5 Higgs bosons are left, 2 of them neutral and CP-even (named h and H^0), 2 of them charged (named $H^\pm$) and one neutral CP-odd boson named A [10].

The supersymmetric partners of the electroweak gauge bosons B and $W^{\pm,0}$, binos and winos, mix with the Higgsinos, the superpartners of the Higgs bosons, since they are all spin-$\frac{1}{2}$ particles. The corresponding mass eigenstates $\tilde{\chi}_i^0$ $(i = 1, \ldots, 4)$ and $\tilde{\chi}_i^\pm$ $(i = 1, 2)$ are called neutralinos and charginos, respectively. There are scalar superpartners, squarks $\tilde{q}$ and sleptons $\tilde{l}$, of the left- and right-handed states of the quarks and leptons which have different weak charges. The squark or slepton states mix to the mass eigenstates $\tilde{q}_{1,2}$ and $\tilde{l}_{1,2}$, respectively, where $\tilde{q}_1, \tilde{l}_1$ are the lighter ones by convention. The whole particle content of the MSSM is shown in Fig. 4.1.

---

[1] Supersymmetric theories require at least 2 Higgs doublets, one for the up and one for the down type fermions, since only one would lead to electroweak gauge anomalies [7].

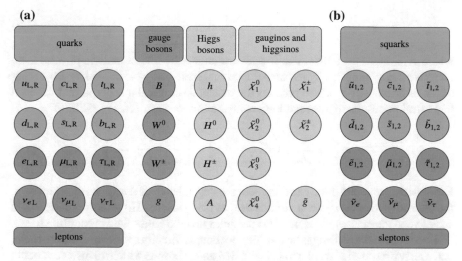

Fig. 4.1 The MSSM particle content consists of the SM particles (**a**) and their superpartners (**b**). **b** shows the mass eigenstates of the SUSY particles

4.2   R-Parity

The generic supersymmetric Lagrangian contains terms which violate the conservation of baryon and lepton number [10]. Since searches for proton decays lead to a very stringent lower bound for the proton lifetime, a new conserved quantum number, called R-parity, is introduced to prevent baryon and lepton number violation and, therefore, proton decay via the exchange of scalar partners of the SM fermions. R-parity is defined as [11]

$$R = (-1)^{2s+3B+L} \quad , \tag{4.4}$$

where s is the spin quantum number and B and L are the baryon and lepton numbers, and is chosen such that $R = +1$ for SM and $R = -1$ for supersymmetric particles. As a consequence, SUSY particles are always produced in even numbers and a sparticle can only decay into an odd number of other SUSY particles. Furthermore, the lightest supersymmetric particle (LSP) has to be stable and cosmological constraints prefer it to be electrically neutral [12].

4.3   The Hierarchy Problem

Since the contributions of bosons and fermions to the Higgs mass corrections have opposite signs, SUSY is an elegant way to address the hierarchy problem [13–16] mentioned in Sect. 2.5. The quantum corrections to the Higgs squared mass parameter m_H^2 from a fermion loop (cf. Fig. 4.2a) are given by [7]

$$\Delta m_H^2 = -\frac{|y_f|^2}{8\pi^2} \Lambda_{\mathrm{UV}}^2 + \cdots \tag{4.5}$$

where Λ_{UV} is the ultraviolet cutoff scale assuming the SM of particle physics is an effective theory replaced by its extension at that scale and y_f is the Yukawa coupling of the fermion. SUSY, which predicts the existence of a scalar superpartner for each SM fermion, results in a second loop contribution to Δm_H^2 (cf. Fig. 4.2b) of

$$\Delta m_H^2 = \frac{|y_S|^2}{16\pi^2} \left(\Lambda_{\mathrm{UV}}^2 - 2m_S^2 \ln\left(\frac{\Lambda_{\mathrm{UV}}}{m_S}\right) + \cdots \right) \tag{4.6}$$

Dimensional regularisation can eliminate the dependence of Δm_H^2 on Λ_{UV}^2, however, the dependence of Δm_H^2 on m_S^2 indicates that the measured Higgs boson mass m_H is sensitive to the masses of the heaviest particles that the Higgs field is coupling to [7].

Even in case the SM Higgs boson does not couple directly to a beyond-SM particle, e.g. a heavy fermion F that interacts with SM gauge bosons which themselves couple to the Higgs boson (cf. Fig. 4.2c and d), the quantum corrections to the Higgs squared mass parameter is given by [7]

$$\Delta m_H^2 = C_H T_F \left(\frac{g_F^2}{16\pi^2}\right)^2 \left(a\Lambda_{\mathrm{UV}}^2 + 24m_F^2 \ln\left(\frac{\Lambda_{\mathrm{UV}}}{m_F}\right) + \cdots \right) \tag{4.7}$$

where C_H and T_F are the quadratic Casimir invariant and the Dynkin index of the respective gauge groups and g_F is the appropriate gauge coupling. As for Eq. (4.6), the Higgs boson mass is sensitive to m_F.

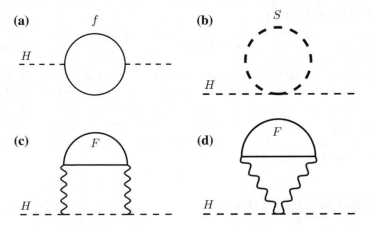

Fig. 4.2 One- and two-loop quantum corrections to the Higgs squared mass parameter m_H^2 due to **a** a Dirac fermion f, **b** a scalar S, **c** and **d** a heavy fermion F that couples only indirectly to the Standard Model Higgs through gauge interactions [7]

Unbroken SUSY would imply $y_S = |y_f|^2$ and thus, nicely cancel all contributions. However, since on the electroweak scale, SUSY is already a broken symmetry, the superpartners of the fermions with large Yukawa couplings $|y_f|$, i.e. the squarks of the third generation, are expected to be light in order to avoid unnatural *fine-tuning* of the Higgs boson mass. The superpartners of the left- and right-handed top quarks, $\tilde{t}_L$ and $\tilde{t}_R$, respectively, mix to form the mass eigenstates $\tilde{t}_1$ and $\tilde{t}_2$, where $\tilde{t}_1$ is defined to have the smaller mass. To serve as a solution to the Hierarchy problem and explain the measured Higgs boson mass of 125 GeV, the top squark mass eigenstates are usually significantly non-degenerated [17] resulting in the mass of the lighter top squark $\tilde{t}_1$ being accessible for the experiments at the Large Hadron Collider.

4.4 Supersymmetric Dark Matter

As stated in Sect. 4.2, the LSP is a stable particle. In addition, it has to be electrically neutral and cannot carry any colour charge [11, 18]. Thus, the LSP brings all basic requirements of a Dark Matter candidate. One possible LSP could be the sneutrino. However, to match the relic density, the sneutrino mass has to be in the range of 550–2300 GeV which is excluded by direct Dark Matter detection experiments [19]. Gravitinos, the supersymmetric partners of a potential spin-2 graviton which is the quantum field mediating the gravitational interaction in extensions of the SM, can be the LSP in gauge mediated SUSY [20]. Although gravitinos are theoretically strongly motivated to contribute to Dark Matter, they cannot be detected by conventional Dark Matter searches since they only interact gravitationally [21].

The most promising supersymmetric Dark Matter candidate is the lightest neutralino $\tilde{\chi}_1^0$. It fulfils all properties of a WIMP and its mass can be of the order $\mathcal{O}(100\,\text{GeV})$ which allows its production at the Large Hadron Collider.

4.5 Supersymmetry and the Open Questions of Particle Physics

The reason why Supersymmetry clearly belongs to the most compelling extensions of the SM of particle physics is not only that it elegantly solves the Hierarchy problem and provides an excellent Dark Matter candidate. SUSY can explain the asymmetry between matter and antimatter [22], enable the gauge coupling unification at the Grand Unification scale [23] and even solve the strong CP problem [24]. Figure 4.3 visualises the great theoretical success of Supersymmetry. However, no supersymmetric particle has been discovered yet. A search for R-parity conserving SUSY, i.e. the lighter top squark $\tilde{t}_1$ will be presented in Chap. 11.

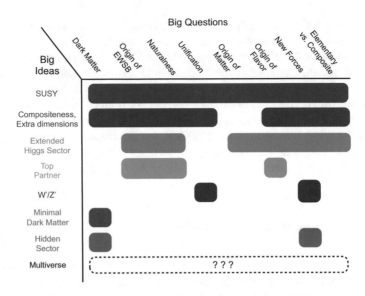

Fig. 4.3 Overlap between the questions of modern particle physics and ideas about their solutions [25]

References

1. Golfand YuA, Likhtman EP (1971) Extension of the algebra of Poincare group generators and violation of p invariance. JETP Lett 13:323–326
2. Volkov DV, Akulov VP (1973) Is the neutrino a goldstone particle? Phys Lett B 46:109–110. https://doi.org/10.1016/0370-2693(73)90490-5
3. Wess J, Zumino B (1974) Supergauge transformations in four dimensions. Nucl Phys B 70:39–50. https://doi.org/10.1016/0550-3213(74)90355-1
4. Wess J, Zumino B (1974) Supergauge invariant extension of quantum electrodynamics. Nucl Phys B 78:1. https://doi.org/10.1016/0550-3213(74)90112-6
5. Ferrara S, Zumino B (1974) Supergauge invariant Yang-Mills theories. Nucl Phys B 79:413. https://doi.org/10.1016/0550-3213(74)90559-8
6. Salam A, Strathdee JA (1974) Super-symmetry and non-Abelian gauges. Phys Lett B 51:353–355. https://doi.org/10.1016/0370-2693(74)90226-3
7. Martin SP (2010) A supersymmetry primer. Adv Ser Direct (hep-ph) 21:1–153. https://doi.org/10.1142/9789814307505_0001
8. Fayet P (1976) Supersymmetry and weak, electromagnetic and strong interactions. Phys Lett B 64:159. https://doi.org/10.1016/0370-2693(76)90319-1
9. Fayet P (1977) Spontaneously broken supersymmetric theories of weak, electromagnetic and strong interactions. Phys Lett B 69:489. https://doi.org/10.1016/0370-2693(77)90852-8
10. Bagger JA (1996) Weak scale supersymmetry: theory and practice. arXiv: hep-ph/9604232 [hep-ph]
11. Djouadi A et al (1998) The minimal supersymmetric standard model: group summary report
12. Ellis JR et al (1984) Supersymmetric relics from the big bang. Nucl Phys B 238:453–476. https://doi.org/10.1016/0550-3213(84)90461-9
13. Dimopoulos S, Georgi H (1981) Softly broken supersymmetry and SU(5). Nucl Phys B 193:150. https://doi.org/10.1016/0550-3213(81)90522-8
14. Sakai N (1981) Naturalness in supersymmetric guts. Z Phys C 11:153. https://doi.org/10.1007/BF01573998

15. Dimopoulos S, Raby S, Wilczek F (1981) Supersymmetry and the scale of unification. Phys Rev D 24:1681–1683. https://doi.org/10.1103/PhysRevD.24.1681
16. Ibanez LE, Ross GG (1981) Low-energy predictions in supersymmetric grand unified theories. Phys Lett B 105:439. https://doi.org/10.1016/0370-2693(81)91200-4
17. Demir DA, Ün, CS (2014) Stop on top: SUSY parameter regions, fine-tuning constraints. Phys Rev D 90:095015. https://doi.org/10.1103/PhysRevD.90.095015
18. Bertone G, Hooper D, Silk J (2005) Particle dark matter: evidence, candidates and constraints. Phys Rep 405:279–390. https://doi.org/10.1016/j.physrep.2004.08.031
19. Falk T, Olive KA, Srednicki M (1994) Heavy sneutrinos as dark matter. Phys Lett B 339:248–251. https://doi.org/10.1016/0370-2693(94)90639-4
20. Giudice GF, Rattazzi R (1999) Theories with gauge mediated supersymmetry breaking. Phys Rep 322:419–499. https://doi.org/10.1016/S0370-1573(99)00042-3
21. Feng JL, Rajaraman A, Takayama F (2003) Superweakly interacting massive particles. Phys Rev Lett 91:011302. https://doi.org/10.1103/PhysRevLett.91.011302
22. Dine M, Kusenko A (2003) The origin of the matter—antimatter asymmetry. Rev Mod Phys 76:1. https://doi.org/10.1103/RevModPhys.76.1
23. de Boer W (1994) Grand unified theories and supersymmetry in particle physics and cosmology. Prog Part Nucl Phys 33:201–302. https://doi.org/10.1016/0146-6410(94)90045-0
24. Ravi K (1996) Solution to the strong CP problem: supersymmetry with parity. Phys Rev Lett 76:3486–3489. https://doi.org/10.1103/PhysRevLett.76.3486
25. Gershtein Y et al (2013) Working group report: new particles, forces, and dimensions. In: Proceedings, 2013 community summer study on the future of U.S. particle physics: snowmass on the Mississippi (CSS2013): Minneapolis, MN, USA, 29 July–6 August 2013. arXiv: 1311.0299 [hep-ex]

Part II
The Experimental Setup

Chapter 5
The Large Hadron Collider at CERN

According to Louis de Broglie's famous hypothesis of massive particles having a quantum mechanical wavelength $\lambda \propto 1/p$ [1], for probing elementary particles and their interactions, the highest possible energies are required. To sustain reproducibility and maximise the statistics of the probes, particle accelerators are used. Currently, the world's most powerful particle accelerator is the Large Hadron Collider (LHC) at CERN described in this chapter. There is also the possibility of probing high-energy particles accelerated by astrophysical sources, but their occurrence on Earth is much more rare and delocalised than when they are produced in the laboratory. Cosmically accelerated particles are beyond the scope of this document but are discussed in several textbooks [2, 3].

The detection of particle collisions at very high energies and collision rates pose the main challenges for current particle detectors. The LHC hosts, among others, two multi-purpose experiments built to cope with these conditions, namely ATLAS and CMS; the former will be introduced in Chap. 6.

Only the precise measurement of the collisions and all decay products involved allows for a profound understanding of particle physics at high energies. The momenta of charged particles are measured using tracking detectors inside a magnetic field and energy deposits of particles stopped inside the detector are estimated using calorimeters. A more detailed description of the reconstruction of muons inside ATLAS including its performance is given in Chap. 7.

The LHC is a two-ring-superconducting-hadron accelerator [4] which is part of the CERN accelerator complex. It is built in a tunnel with a circumference of 26.7 km where the Large Electron-Positron collider (LEP) [5, 6] was housed previously. Since the tunnel was originally designed for an electron-positron collider, it has eight ~530 m long straight sections meant to host radio frequency cavities to compensate the high synchrotron radiation losses of LEP, complemented with eight interaction regions in between (cf. Fig. 5.1). While the caverns in the interaction regions 2, 4, 6 and 8 were hosting the LEP experiments already, two new caverns were constructed around the interaction points 1 (Point 1 or IP1) and Point 5 for the ATLAS [7] and CMS [8] experiments, respectively. The caverns in the interaction regions 2 and 8

© Springer Nature Switzerland AG 2019

N. M. Köhler, *Searches for the Supersymmetric Partner of the Top Quark, Dark Matter and Dark Energy at the ATLAS Experiment*, Springer Theses, https://doi.org/10.1007/978-3-030-25988-4_5

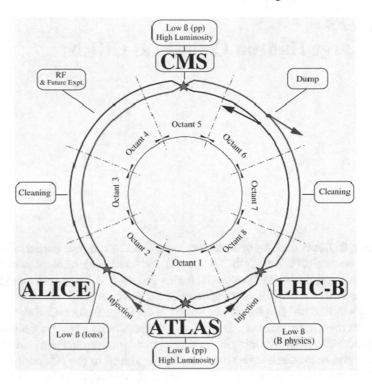

Fig. 5.1 Schematic layout of the LHC tunnel [4]

are hosting the ALICE [9] and LHCb [10] experiments, while for the remaining four interaction regions, the beam crossings are suppressed to prevent unnecessary disruption of the beams [11]. Octant 3 and 7 each host two beam collimation systems while two radio-frequency cavities are installed within the 4th octant. In the case of refilling the LHC or emergencies, the beams are dumped using so-called *kicker* magnets [12] in interaction region 6 [4].

Contrary to LEP, the high beam intensity required to achieve high interaction rates does not allow for the usage of anti-proton beams. Therefore, the LHC consists of two rings with counter rotating proton beams which require the presence of two inverse magnetic fields. Due to the relatively small 3.7 m diameter of the LEP tunnel, the LHC needs to employ 1232 superconducting state-of-the-art dipole magnets operating at temperatures of 1.9 K [13] which are able to create peak magnetic fields up to 8.33 T [12].

Like LEP, the LHC relies on the pre-acceleration chain of the CERN accelerator complex (cf. Fig. 5.2). For proton-proton (*pp*) collisions, after stripping of the electrons of $\sim 1.15 \cdot 10^{11}$ hydrogen atoms, the so-called proton *bunch* is passing a linear accelerator (Linac 2) which increases the proton energy up to 50 MeV. Then, a circular accelerator of 160 m in circumference, the *Proton Synchrotron Booster* accelerates the protons to 1.4 GeV, before the *Proton Synchrotron* (630 m in cir-

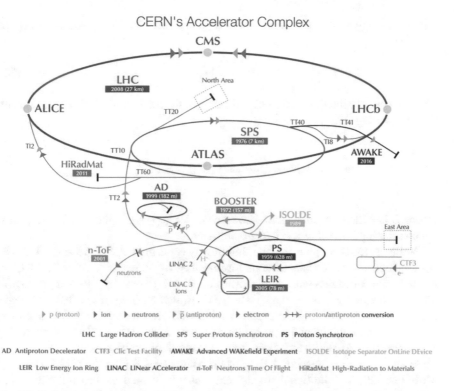

CERN's Accelerator Complex

Fig. 5.2 Before injection into the LHC, the protons undergo several pre-acceleration steps. In a linear accelerator (*Linac 2* for protons, *Linac 3* for $^{208}_{82}$Pb), they obtain an energy of 50 MeV and are injected into the *Proton Synchrotron Booster* (PSB) where they are accelerated to 1.4 GeV. The *Proton Synchrotron* (PS) and the *Super Proton Synchrotron* (SPS) increase the energy to 25 GeV and 450 GeV, respectively. At the last energy, the proton bunches are injected into the LHC ring [17]

cumference) and the *Super Proton Synchrotron* (7 km in circumference) increase the energy to 25 GeV and 450 GeV, respectively [14]. Last, the bunches are brought to the LHC injection points locates ~150 m left of IP2 and ~160 m right of IP8 by transfer lines.

The figure of merit for colliders such as the LHC is the *instantaneous luminosity* [14]

$$\mathscr{L} = \frac{N_b^2 n_b f_{\text{rev}} \gamma}{4\pi \epsilon_n \beta^*} \quad , \tag{5.1}$$

where N_b is the number of particles per bunch, n_b the number of bunches per beam, f_{rev} the revolution frequency telling the number of circulations of each bunch per second, and γ the relativistic gamma factor. In the denominator, the product of ϵ_n, the normalised transverse beam emittance and β^*, the beta function at the interaction point, gives a measure of the cross-sectional size of the bunch [15]. The instantaneous luminosity is related to the event rate of a physics process having the cross section σ by [16]

$$\dot{N} = \mathscr{L} \cdot \sigma \quad . \tag{5.2}$$

Thus, the integration of Eq. (5.2) over time gives the total number of expected events for a measuring time T

$$N = L \cdot \sigma = \int_0^T \mathscr{L}(t) \cdot \sigma \, dt \quad , \tag{5.3}$$

where L is called the *integrated luminosity*.[1] Here, *event* rather denotes a proton bunch crossing instead of a single *pp* collision.

While the LHC supplies the high luminosity experiments ATLAS and CMS with *pp* collisions at peak instantaneous luminosities of $\mathscr{L} = 10^{34} \, \mathrm{cm}^{-2} \, \mathrm{s}^{-1}$, LHCb is limited to peak instantaneous luminosities of $\mathscr{L} = 10^{32} \, \mathrm{cm}^{-2} \, \mathrm{s}^{-1}$ to study decays of hadrons containing *b* or *c* quarks. The LHC is also capable of accelerating heavy ions ($^{208}_{82}\mathrm{Pb}$) and providing either lead-lead or lead-proton collisions with an energy of 2.8 TeV per nucleon and a peak luminosity of $\mathscr{L} = 10^{27} \, \mathrm{cm}^{-2} \, \mathrm{s}^{-1}$, the working environment of ALICE which is studying the quark gluon plasma [4].

Apart from the four big LHC experiments, LHCf [18], TOTEM [19] and MoEDAL [20] are additional smaller experiments located next to the main interaction points.

5.1   Proton–Proton Collisions at the LHC

The LHC design value for the centre-of-mass energy of *pp* collisions is $\sqrt{s} = 14$ TeV at an instantaneous luminosity of $\mathscr{L} = 10^{34} \, \mathrm{cm}^{-2} \, \mathrm{s}^{-1}$. Being operated with $1.15 \cdot 10^{11}$ protons per bunch and 2808 proton bunches per beam, the LHC reaches a revolution frequency of 40 MHz [4]. However, in order to smoothly run and understand the accelerator, in 2009, the first *pp* collisions at $\sqrt{s} = 900$ GeV and $\mathscr{L} = 3 \cdot 10^{26} \, \mathrm{cm}^{-2} \, \mathrm{s}^{-1}$ took place before the energy was increased to $\sqrt{s} = 7$ TeV in 2010. In 2012, the energy was increased again to $\sqrt{s} = 8$ TeV and peak luminosities of about $\mathscr{L} = 8 \cdot 10^{33} \, \mathrm{cm}^{-2} \, \mathrm{s}^{-1}$ were reached. In February 2013, the LHC was shut down for maintenance work and an energy upgrade to $\sqrt{s} = 13$ TeV at which the collider continued data-taking in May 2015. From 2015 to 2017, the LHC delivered 93fb^{-1} of *pp* collision data at $\sqrt{s} = 13$ TeV (cf. Fig. 5.3a). In October 2017, for the first time, the LHC doubled its design luminosity and measured an instan-

---

[1]Note, that the instantaneous luminosity in the LHC is not constant over time, but decays due to the degradation of intensities and emittances of the circulating beams, mainly due to beam loss from collisions.

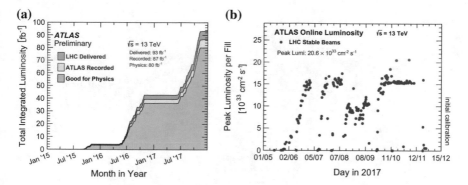

Fig. 5.3 ATLAS luminosity summary plots for pp collision data taking at $\sqrt{s} = 13$ TeV during stable beams. **a** shows the cumulative luminosity versus time delivered to ATLAS (green), recorded by ATLAS (yellow), and certified to be good quality data (blue) in 2015–2017, **b** shows the peak instantaneous luminosity delivered to ATLAS for each LHC fill as a function of time in 2017 [21]

taneous peak luminosity of $\mathscr{L} = 2 \cdot 10^{34}\,\mathrm{cm}^{-2}\,\mathrm{s}^{-1}$, as depicted in Fig. 5.3b. These unprecedented peak luminosities pose severe challenges for both accelerators and detectors. For the latter, the main challenges are radiation damage and the occurrence of multiple pp collisions per bunch crossing, commonly referred to as *pile-up* (cf. Sect. 6.7). Some of these challenges for the precision muon chambers of the ATLAS detector will be discussed in Chap. 9.

Phenomenology of Proton–Proton Collisions

Since protons are composite particles consisting of three *valence* quarks and an arbitrary number of gluons and *sea* quarks, commonly referred to as *partons*, a pp collision depends on the particular dynamics of the compound proton system. Depending on the momentum transfer between the constituents of the protons, a pp scattering process can be classified into a *hard* and a *soft scattering* process [22]. Usually, a hard scatter can be precisely predicted by perturbative QCD calculations while for the soft scatter, phenomenological models have to be used. The soft scatter of the proton remnants is also referred to as the *underlying event* and has been intensively studied for proton-antiproton collisions at the Tevatron [23]. The phenomenological models are tuned to the collision data by so-called *underlying event tunes* [24].

According to the factorisation theorem of QCD [25], the cross-section of two protons interacting to an arbitrary final state X, $\sigma(p_a p_b \to X)$, can be factorised into the hard scattering cross-section of two partons a and b interacting and producing X convoluted with the *parton distribution functions* (p.d.f.s) f_a and f_b in the protons

$$\sigma(p_a p_b \to X) = \sum_{a \in p_a, b \in p_b} \int_0^1 dx_a \int_0^1 dx_b \, f_a(x_a, \mu_F^2) \, f_b(x_b, \mu_F^2) \cdot \sigma_{ab \to X}(x_a P_a, x_b P_b, \mu_R^2) \quad .$$

$$(5.4)$$

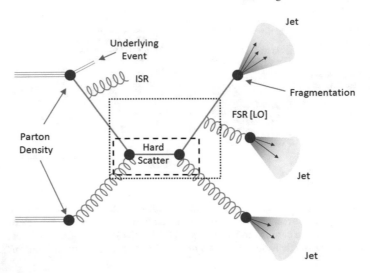

Fig. 5.4 Schematic layout of a proton-proton collision [30]. Note that a calculation in leading order of the hard scatter (dashed line) will assign a jet to final state radiation that, in a next-to-leading order calculation (dotted line), would be included in the hard scatter itself

x_a and x_b ($\in [0, 1]$) are the momentum fractions of the partons a and b inside the protons p_a and p_b, respectively, μ_F is the factorisation scale describing the boundary between hard and soft scatter and μ_R the renormalisation scale resulting from the perturbative calculation of the parton-level cross-section. Inside the partonic cross-section $\sigma_{ab \to X}$, P_a and P_b are the momenta of the proton p_a and p_b respectively. Figure 5.4 sketches the separation into hard and soft scatter of a pp collision as well as the underlying event. The parton distribution functions needed for the cross-section calculations cannot be derived from first principles but have been extensively estimated based on various experimental measurements [26–29].

5.2   The Worldwide LHC Computing Grid

The average file size of one single raw event recorded by ATLAS is roughly 1.6 Megabyte (MB) [31]. Since the revolution frequency of the LHC is 40 MHz, recording every collision would result in a data rate of ~60 Terabyte (TB) per second which makes it impossible to read out and store all events with current computing technologies. By applying a sophisticated triggering system (cf. Sect. 6.5), the event rate can be reduced to 1 kHz. However, the fact that the LHC hosts four main experiments which have roughly the same computing requirements paired with the fact that the experimental data needs to be provided to approximately 5000 scientist working distributed all over the world, led to the construction of the Worldwide LHC Computing Grid (WLCG) [32]. It is designed to cope with the requirement of handling around 15 Petabyte (15,000 TB) of collision data per year including its redundant preservation over a 15-year lifetime and its availability at any time for all collaborators.

The WLCG is a distributed computing grid of more than 160 computing centres in 42 countries, linking up national and international grid infrastructures. The distributed layout provides the advantages of easier and less cost-intensive of maintenances and upgrades as well as a smaller sensitivity to single points of failure or local incidents. The WLCG is hierarchically structured into different tiers:

- A Tier-0 centre is located at CERN which is connected by three 100 Gbit/s data links to a second Tier-0 centre located at Budapest, Hungary. At the Tier-0 centres, the first-pass reconstruction takes place. Then, a copy of the data gets distributed to a Tier-1 centre for storage and further processing.
- Tier-1 centres are responsible for the permanent storage of all raw, processed and simulated data. Currently, there are 14 active Tier-1 centres around the world [33].
- The task of Tier-2 centres is to provide computational capacity for end-user analysis and Monte Carlo event simulations. Data generated at Tier-2 centres meant for longer term storage are sent back to Tier-1 centres. Currently, there are 149 active Tier-2 centres around the world [33].
- Local computing farms at universities and other institutes are commonly referred to as Tier-3. They are not managed by the WLCG but have to be provided with access to the respective WLCG sites.

To protect the WLCG resources from outside attacks but also ensure their availability to meet the scientific aims of the project, a *Virtual Organisation Membership Service* is employed requiring the participating institutes to define and agree on robust security policies, procedures and guides enabling the building and maintenance of *trust* between the various bodies involved [32].

References

1. De Broglie M (1921) Les phénoménes photo-électriques pour les rayons X et les spectres corpusculaires des éléments. J Phys Radium 2:9. https://doi.org/10.1051/jphys-rad:0192100209026500
2. Gaisser TK, Engel R, Resconi E (2016) Cosmic rays and particle physics. Cambridge University Press. isbn: 9781316598436
3. Lemoine M, Sigl G (2001) Physics and astrophysics of ultra high energy cosmic rays. Lecture notes in physics. Springer, Berlin Heidelberg. isbn: 9783540428992
4. Evans L, Bryant P (2008) LHC machine. J Instrum 3(08):S08001
5. LEP design report, vol 1. The LEP injector chain (1983). http://cds.cern.ch/record/98881
6. LEP design report: vol 2. The LEP main ring (1984). http://cds.cern.ch/record/102083
7. ATLAS Collaboration (2008) The ATLAS experiment at the CERN Large Hadron Collider. JINST 3:S08003. https://doi.org/10.1088/1748-0221/3/08/S08003
8. CMS Collaboration (2008) The CMS experiment at the CERN LHC. JINST 3:S08004. https://doi.org/10.1088/1748-0221/3/08/S08004

9. Aamodt K et al (2008) The ALICE experiment at the CERN LHC. JINST 3:S08002. https://doi.org/10.1088/1748-0221/3/08/S08002
10. Alves A et al (2008) The LHCb detector at the LHC. JINST 3:S08005. https://doi.org/10.1088/1748-0221/3/08/S08005
11. Bruning O et al (2004) LHC design report: vol 2. The LHC infrastructure and general services. http://cds.cern.ch/record/815187
12. Bruning OS et al (2004) LHC design report: vol 1. The LHC main ring. http://cds.cern.ch/record/782076
13. Casas J et al (1992) Design concept and first experimental validation of the superfluid helium system for the Large Hadron Collider (LHC) project at CERN. In: Proceedings of the fourteenth international cryogenic engineering conference and international cryogenic materials conference cryogenic engineering & superconductor technology. Cryogenics, vol 32, pp 118–121. issn: 0011-2275. https://doi.org/10.1016/0011-2275(92)90122-Q
14. Benedikt M et al (2004) LHC design report: vol 3. The LHC injector chain. http://cds.cern.ch/record/823808
15. Wiedemann H (1999) Particle accelerator physics I, vol 1. Springer. isbn: 9783540646716. https://books.google.de/books?id=nTJOUx5oQQ0C
16. Frauenfelder H, Henley EM, Reck M (1999) Teilchen und Kerne: dieWelt der subatomaren Physik. Oldenbourg. isbn: 9783486244175
17. Haffner J (2013) The CERN accelerator complex. https://cds.cern.ch/record/1621894
18. Adriani O et al (2008) The LHCf detector at the CERN Large Hadron Collider. JINST 3:S08006. https://doi.org/10.1088/1748-0221/3/08/S08006
19. Anelli G et al (2008) The TOTEM experiment at the CERN Large Hadron Collider. JINST 3:S08007. https://doi.org/10.1088/1748-0221/3/08/S08007
20. Pinfold J et al (2009) Technical design report of the MoEDAL experiment. CERN-LHCC-2009-006
21. ATLAS Collaboration (2018). https://twiki.cern.ch/twiki/bin/view/AtlasPublic/LuminosityPublicResultsRun2
22. Campbell JM, Huston JW, Stirling WJ (2007) Hard interactions of quarks and gluons: a primer for LHC physics. Rep Prog Phys 70:89. https://doi.org/10.1088/0034-4885/70/1/R02
23. Field RD (2001) The underlying event in hard scattering processes. In: eConf C010630, p P501. arXiv: hep-ph/0201192 [hep-ph]
24. Moraes A, Buttar C, Dawson I (2007) Prediction for minimum bias and the underlying event at LHC energies. Eur Phys J C 50:435–466. https://doi.org/10.1140/epjc/s10052-007-0239-1
25. Collins JC, Soper DE, Sterman GF (1989) Factorization of hard processes in QCD. Adv Ser Direct High Energy Phys 5:1–91. https://doi.org/10.1142/9789814503266_0001
26. Martin AD et al (2009) Parton distributions for the LHC. Eur Phys J C 63:189–285. https://doi.org/10.1140/epjc/s10052-009-1072-5
27. Pumplin J et al (2002) New generation of parton distributions with uncertainties from global QCD analysis. JHEP 07:012. https://doi.org/10.1088/1126-6708/2002/07/012
28. Ball RD et al (2013) Parton distributions with LHC data. Nucl Phys B 867:244–289. https://doi.org/10.1016/j.nuclphysb.2012.10.003
29. Ball RD et al (2011) Impact of heavy quark masses on parton distributions and LHC phenomenology. Nucl Phys B 849:296–363. https://doi.org/10.1016/j.nuclphysb.2011.03.021
30. Bhatti A, Lincoln D (2010) Jet physics at the tevatron. Annu Rev Nucl Part Sci 60:267–297. https://doi.org/10.1146/annurev.nucl.012809.104430
31. Duckeck G et al (2005) ATLAS computing: technical design report
32. Bird I et al (2005) LHC computing grid. Technical design report
33. Worldwide LHC computing grid (2018). http://wlcg.web.cern.ch/

Chapter 6
The ATLAS Detector

ATLAS (A Toroidal LHC ApparatuS) [1] is a multi-purpose detector for both SM precision measurements and the search for new physics phenomena at the TeV energy scale. The main first goal of the ATLAS physics program was the search for the Higgs boson.

ATLAS is located within the interaction region 1 of the LHC about 100 m below ground. With 46 m in length, 25 m in height and a mass of about 7000 tonnes, it is the biggest particle detector built so far (cf. Fig. 6.1). The detector is cylindrically symmetric around the beam axis as well as forward-backward symmetric with respect to the interaction point. The *barrel region* consists of cylindrical layers around the beam axis while the *end-cap regions* contain disk shaped layers.

By convention, the ATLAS coordinate system is right-handed with origin at the interaction point. The z-axis points along the beam axis and the x-axis towards the centre of the LHC. Thus, the x-y-plane is the transverse plane with respect to the beam. The transverse momentum of a particle of four-momentum $p^{\mu} = (E/c, \mathbf{p})$ is defined as $p_{\mathrm{T}} = \sqrt{p_x^2 + p_y^2}$. Each point of the detector can be described by cylindrical coordinates (r, θ, ϕ) where ϕ is the azimuthal angle around the beamline (z-axis) with $\phi = 0$ pointing along the x-axis and θ the polar angle with respect to the z-axis. The *pseudorapidity* η of a particle with momentum $\mathbf{p}$ at the angle θ to the z-axis is defined as [2]

$$\eta = -\ln\left(\tan\left(\frac{\theta}{2}\right)\right) \quad . \tag{6.1}$$

With $|\mathbf{p}| = p$, it can also be expressed by using the longitudinal component p_z of the particle's momentum

$$\eta = \frac{1}{2} \cdot \ln\left(\frac{p + p_z}{p - p_z}\right). \tag{6.2}$$

© Springer Nature Switzerland AG 2019
N. M. Köhler, *Searches for the Supersymmetric Partner of the Top
Quark, Dark Matter and Dark Energy at the ATLAS Experiment*,
Springer Theses, https://doi.org/10.1007/978-3-030-25988-4_6

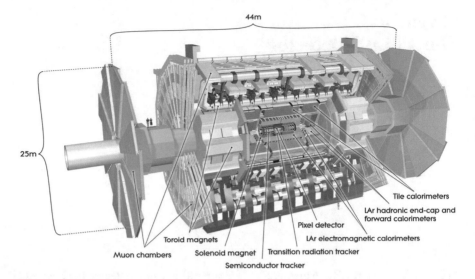

44m

25m

Tile calorimeters

LAr hadronic end-cap and
forward calorimeters

Pixel detector

LAr electromagnetic calorimeters

Toroid magnets

Muon chambers   Solenoid magnet | Transition radiation tracker

Semiconductor tracker

Fig. 6.1 Cut-away view of the ATLAS detector [1]

For ultra-relativistic particles ($p \gg m$), the pseudorapidity coincides with the rapidity y

$$\eta \approx y = \frac{1}{2} \cdot \ln\left(\frac{E + E_z}{E - E_z}\right), \tag{6.3}$$

thus, the differences in the pseudorapidity give a Lorentz invariant measure for boosts along the beam pipe [2]. The convenience of the pseudorapidity compared to the rapidity in high energy applications is that it can be derived directly from geometrical quantities without knowing any particle properties. Therefore, two useful measures for the angular separation between two objects $o_1 = (r_1, \theta_1, \phi_1)$ and $o_2 = (r_2, \theta_2, \phi_2)$ are defined,

$$\Delta R = \sqrt{(\Delta\eta)^2 + (\Delta\phi)^2} = \sqrt{(\eta_2 - \eta_1)^2 + (\phi_2 - \phi_1)^2} \quad \text{and} \quad (6.4)$$

$$\Delta R_y = \sqrt{(\Delta y)^2 + (\Delta\phi)^2} = \sqrt{(y_2 - y_1)^2 + (\phi_2 - \phi_1)^2}. \tag{6.5}$$

Since the ATLAS detector is supposed to record pp collision data at high energy and luminosity, several requirements have to be met:

- High momentum resolution and reconstruction efficiency for charged particles.
- Large pseudorapidity and azimuthal angle coverage and high detector granularity.
- Measurements of the energy of photons, electrons and jets and of the missing transverse energy with high accuracy.
- An efficient trigger system to cope with the high event rates.

To satisfy these demands, the detector consists of four sub-detectors arranged in an onion like structure around the interaction point.

6.1 The Inner Detector

The tracks of charged particles curved in a 2 T magnetic field produced by a super-conducting solenoid are measured by the Inner Detector (ID) determining both their direction and momentum. The ID consists of 3 parts (cf. Fig. 6.2): The Silicon Pixel Detector, the Semi-Conductor Tracker (SCT) and the Transition Radiation Tracker (TRT).

The Silicon Pixel Detector is used for precise primary and secondary vertex measurements. It has 4 concentric layers of silicon pixel sensors in the barrel region and 3 discs perpendicular to the beam axis in the end-cap regions. The innermost pixel layer was inserted after LHC Run 1 and is called *insertable B-layer* [3, 4]. It is only 3.3 cm away from the beam axis and improves the quality of impact parameter reconstruction for tracks used for vertexing and identifying tracks of hadrons containing *b* quarks. The insertable B-layer is surrounded by the former B-layer and two more pixel layers for charged particle tracking. The pixel detector's spatial resolution is 10 μm in the transverse plane and 115 μm in beam direction. It has approximately 86.4 million readout channels.

The Pixel Detector is surrounded by the SCT which consists of 4 layers of silicon microstrip sensors in the barrel region and 9 disks in the end-cap regions to allow for precise track reconstruction in a high track density environment. Its spatial resolution is 17 μm in the transverse plane and 580 μm parallel to the beam axis. Both Silicon Pixel Detector and SCT cover the pseudo rapidity region of $|\eta| < 2.5$.

The SCT is surrounded by the TRT Tracker which is made of 4 mm diameter straw drift tubes consisting of 35 μm thick Kapton with 31 μm diameter gold-plated tungsten-rhenium anode wires. The straw drift tubes are filled with a Xe (70%), CO_2 (27%) and O_2 (3%) gas mixture. The tubes measure charged particles ionising

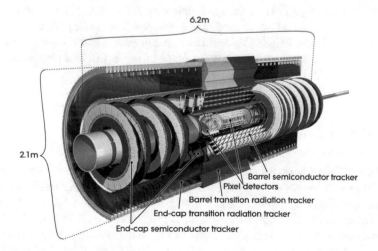

Fig. 6.2 The ATLAS inner detector [1]

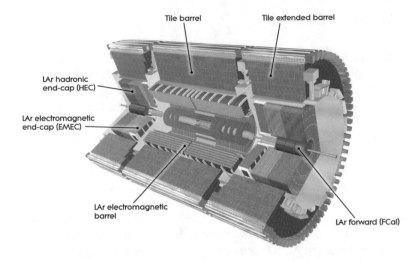

Tile barrel   Tile extended barrel

LAr hadronic
end-cap (HEC)

LAr electromagnetic
end-cap (EMEC)

LAr electromagnetic
barrel

LAr forward (FCal)

Fig. 6.3 The ATLAS calorimeter system [1]

the gas and transition radiation photons in order to identify electrons. The track measurements are only performed in the transverse plane over a pseudo rapidity range of $|\eta| < 2.0$ and with an accuracy of $130\,\mu\mathrm{m}$ per straw.

6.2   The Electromagnetic Calorimeter

The Electromagnetic Calorimeter (cf. Fig. 6.3) measures the energy of electromagnetically interacting particles with high granularity. It consists of lead absorber plates with liquid argon (LAr) as active medium in between. The barrel part has a thickness of about 22 radiation lengths X_0 and the end-cap part is more than 24 X_0 thick. The calorimeter covers a pseudo rapidity range of $|\eta| < 4.9$. It is segmented into 3 longitudinal sections for $|\eta| < 2.5$ and into 2 for $|\eta| > 2.5$.

6.3   The Hadron Calorimeter

The Hadron Calorimeter measures the energy deposits of hadrons produced in the pp collisions. It consists of 3 parts (cf. Fig. 6.3): The Tile Calorimeter ($|\eta| < 1.7$), the Hadronic End-Cap liquid Argon Calorimeter (HEC, $1.7 < |\eta| < 3.2$) and the liquid Argon Forward Calorimeter (FCal, $3.1 < |\eta| < 4.9$). The Tile Calorimeter surrounds the Electromagnetic Calorimeter and is made of scintillating tiles as active medium and steel as absorber. At $\eta = 0$, the total thickness of the Tile Calorimeter is 9.7 radiation lengths. The HEC is located behind the Electromagnetic end-cap

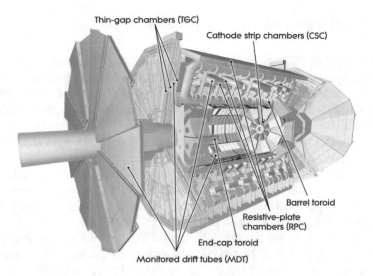

Fig. 6.4 The ATLAS muon spectrometer [1]

Calorimeter and uses liquid argon as active medium and copper as absorber. The FCal has a thickness of around 10 radiation lengths and uses liquid argon as active medium, copper absorber in the innermost of its 3 modules and tungsten absorber in the other two.

6.4 The Muon Spectrometer

The ATLAS Muon Spectrometer (MS, cf. Fig. 6.4) has two main purposes: It provides the first level muon trigger with dedicated trigger chambers (cf. Sect. 6.5) and measures the deflection of the muon trajectories in a toroidal magnetic field using 3 layers of high-precision tracking chambers. The MS is permeated by a 0.3–1.2 T toroidal magnetic field produced by superconducting air-core toroid magnets.

The precision tracking chambers are the Monitored Drift Tube (MDT) chambers which cover a pseudorapidity range of $|\eta| < 2.7$. At large pseudorapidities ($|\eta| > 2.0$), a layer of Cathode Strip Chambers (CSC) is installed in front of the end-cap toroid to cope with the high background rates close to the beam pipe.

The muon trigger chambers are Resistive Plate Chambers (RPC) in the barrel region ($|\eta| < 1.05$) and Thin Gap Chambers (TGC) in the end-cap regions covering a pseudo rapidity range of $|\eta| < 2.4$. The trigger chambers have worse spatial resolution than the MDT chambers but provide much faster signals required for the first level muon trigger.

6.5   The Trigger and Data Acquisition System

Since the nominal LHC bunch crossing rate of 40 MHz does not allow the data acquisition system to continuously readout the full detector information, a dedicated trigger system has to be applied. It has to select events of interest at a recording rate of approximately 1 kHz [5]. The ATLAS Trigger and Data Acquisition (TDAQ) [6] system consists of a hardware-based first level trigger (L1) which reduces the event rate from 40 MHz to 100 kHz and a software-based high-level trigger (HLT) with an average output rate of 1 kHz as depicted in Fig. 6.5.

The L1 trigger consists of the L1 calorimeter (L1Calo) and L1 muon (L1Muon) triggers as well as other subsystems described in [7] which provide inputs to the L1 Central Trigger commissioned during 2016 [8]. The L1 Central Trigger consists of the Central Trigger Processor (CTP), the Muon-to-CTP interface (MUCPTI) and the L1 topological trigger (L1Topo). The MUCPTI receives trigger objects from the L1Muon system, the L1Topo trigger performs selections based on geometric or kinematic association between trigger objects received from the L1Calo or L1Muon systems. Both provide inputs to the CTP which forms the final first level trigger decision.

After the L1 trigger decision is made, the events are buffered in the Read-Out System (ROS) while the information on the Region-of-Interest (RoI) is sent to the HLT. The HLT uses tracking information from the ID and the MS as well as finer-granularity calorimeter information. Most HLT triggers use a two-stage approach in order to reduce processing time. In a fast first-pass reconstruction, the majority of the events is rejected while in a slower precision reconstruction, the remaining events are analysed.

The ATLAS trigger menu [5] defines a list of L1 and HLT triggers to cover all signatures relevant to the ATLAS physics program while staying within the limits of storage and bandwidth. It consists of so-called *primary* triggers used for physics analyses, *support* triggers meant for trigger efficiency and performance measurements, *alternative* triggers testing new reconstruction algorithms and *calibration* triggers used for detector calibration by operating at high rates but storing only the relevant information for the calibration.

Ideally, primary triggers are unprescaled, which means all bunch crossings fulfilling the trigger requirements are recorded. However, to stay within the rate constraints for a given LHC luminosity, prescale factors can be applied to both L1 and HLT triggers. The prescale factors can be changed during data-taking in a way that triggers may be disabled or only a certain fraction of the collisions passing the trigger requirements are accepted.

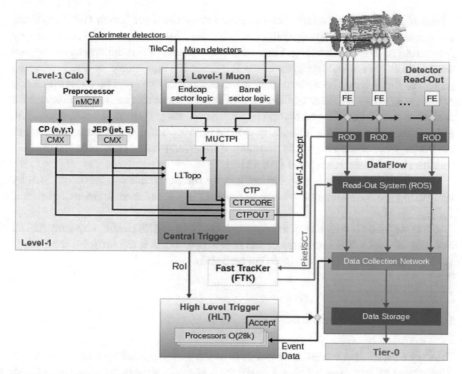

Fig. 6.5 The ATLAS TDAQ system in LHC Run 2 with emphasis on the components relevant for triggering [5]

6.6 Luminosity Measurement

An accurate estimation of the integrated luminosity delivered to ATLAS is one of the main requirements for every physics analysis, i.e. for precision analyses, such as cross-section measurements where the uncertainty on the luminosity measurement is often one of the dominating systematic uncertainties. Following Eq. (5.2), the instantaneous luminosity $\mathscr{L}$ can be expressed by the event rate of inelastic pp collisions $\dot{N}_{\text{inel.}}$ and the corresponding cross-section $\sigma_{\text{inel.}}$ [9]

$$\mathscr{L} = \frac{\dot{N}_{\text{inel.}}}{\sigma_{\text{inel.}}} \ . \tag{6.6}$$

For a storage ring operating at a revolution frequency f_{rev} and having n_b bunches per beam, Eq. (6.6) can be rewritten as

$$\mathscr{L} = \frac{\mu n_b f_{\text{rev}}}{\sigma_{\text{inel.}}} \ , \tag{6.7}$$

where μ is the average number of inelastic interactions per bunch crossing. Thus, the instantaneous and subsequently, the integrated luminosity can be measured by estimating μ during data taking. Since the measurement of μ is depending on the particular choice of detector type, coverage and detection algorithm, the observed value $\mu_{obs.}$ will differ from the real value by the measurement's acceptance and efficiency ϵ, thus, $\mu_{obs.} = \epsilon \cdot \mu$. Therefore, the luminosity measurement has to be calibrated for each detection type. The calibration is performed using dedicated beam-separation scans, where the absolute luminosity can be inferred from direct measurements of the beam parameters, such as the normalised transverse beam emittance and the beta function at the interaction point [10, 11].

A systematic uncertainty on the luminosity measurement can be determined by comparing the measurements of several detector types and measurement algorithms for $\mu_{obs.}$ [12].

The integrated luminosity of the data taken in 2015 and 2016 after requiring that all detector subsystems are operational and the recorded data is uncorrupted is 36.1fb^{-1} with a relative uncertainty estimated to be 3.2%.

6.7   Pile-Up Interactions

Primary vertices are defined as the points in space where pp collisions have occurred. The precise measurement of primary vertices is crucial for the data analysis in ATLAS since the particle trajectories have to be correctly assigned to the hard scatter primary vertex to properly reconstruct the kinematic properties of the event. The superposition of multiple inelastic pp collisions reconstructed in one physics event causes the reconstruction of additional primary vertices apart from the hard scatter, which are usually low-momentum hadronic interactions of other protons inside the same bunch. This effect is referred to as *in-time* pile-up [13]. Since the proton bunches are crossing each 25 ns, it potentially happens that signals from neighbouring bunch crossings can be present simultaneously when the detector is read out, which is called *out-of-time* pile-up. The average number of inelastic interactions per bunch crossing μ (cf. Sect. 6.6) follows a Poisson distribution with mean value $< \mu >$. The value of $< \mu >$ decreases with decreasing beam intensity and increasing emittance during data taking. The mean number of interactions per bunch crossing measured by ATLAS in the years 2015–2017 is shown in Fig. 6.6.

Pile-up is taken into account in simulations by overlaying a varying number of simulated minimum-bias interactions on the hard scatter of the event. Differences between the distribution of the average number of interactions per bunch crossing used in the simulation and the one measured in data are corrected by applying a per-event weight to the simulation reflecting the relative distribution of μ in the statistics of the simulation [13].

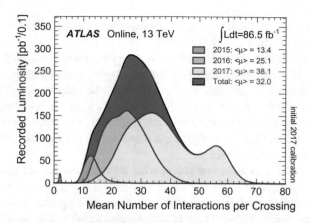

Fig. 6.6 Luminosity-weighted distribution of the mean number of interactions per bunch crossing for pp collision data at $\sqrt{s} = 13\,\mathrm{TeV}$. All data recorded by ATLAS during stable beams is shown. The number of interactions per bunch crossing is calculated from the instantaneous per bunch luminosity $\mathscr{L}_{\mathrm{bunch}}$ as $\mu = \mathscr{L}_{\mathrm{bunch}}\sigma_{\mathrm{inel.}}/f_{\mathrm{rev}}$, where $\sigma_{\mathrm{inel.}}$ is the inelastic pp scattering cross section taken to be 80 mb, and f_{rev} is the LHC revolution frequency. The luminosity shown represents the initial 13 TeV luminosity estimate and includes all 13 TeV pp collision data recorded in 2015–2017 [14]

6.8   Physics Object Identification and Reconstruction

The ID track reconstruction [15] uses the measurements of all three ID parts to reconstruct the momenta and the transverse (d_0) and longitudinal (z_0) impact parameters of the tracks, where d_0 is the smallest distance between the track and the interaction point in the transverse plane and z_0 the z coordinate of the point of closest approach. A vertex reconstruction algorithm [16] uses well reconstructed particle tracks with $p_{\mathrm{T}} > 400\,\mathrm{MeV}$ to determine the primary interaction vertex in a χ^2 fit. Tracks incompatible with the measured vertex by more than 7 standard deviations are used for the reconstruction of another vertex. The vertex with the largest sum of squared transverse momenta of the associated tracks is chosen as the primary vertex of the event, indicating the production of heavy particles in the collision. Events not containing a reconstructed primary vertex are discarded.

Electrons within $|\eta| < 2.47$ are reconstructed from energy deposits in the electromagnetic calorimeter that are matched to tracks in the ID which are required to have a minimum number of hits in the tracking detectors [17]. A likelihood-based analysis algorithm categorises the electron candidates according to *VeryLoose*, *Loose*, *Medium* and *Tight* electron identification criteria [17–19].

The reconstruction of muons with $|\eta| < 2.5$ is first performed independently by track reconstruction in the ID and MS, before the measurements are combined by a global fit of all hits to muon tracks [20]. To reject events with poor fit quality, e.g. for muons originating from decays of charged hadrons, the significance of the charge-over-momentum ratio has to fulfil $\sigma(q/p)/|q/p| < 2$. For the region $2.5 < |\eta| < 2.7$

outside the coverage of the ID, muons are reconstructed in the MS only. At $|\eta| < 0.1$, there are gaps in the MS coverage needed for the cables of the ID. In these regions, muons are identified by ID tracks with energy deposits compatible with a minimum-ionising particle in the calorimeters (calorimeter tagged muons). Muon identification is done by applying quality requirements that suppress background, mainly from pion and kaon decays, while selecting prompt muons with high efficiency and guarantee-ing a robust momentum measurement. The muon identification selections available are *Loose*, *Medium*, *Tight* and *High-p_T* depending on the identification efficiency and background rejection [20].

Events containing at least one muon with a transverse impact parameter $d_0 >$ 0.2 mm and a longitudinal impact parameter $z_0 > 1$ mm are rejected to suppress muons from cosmic rays. For events obtained by a muon trigger selection, the pseudo-rapidity requirements of muons have to be restricted to $|\eta| < 2.4$ due to the coverage of the muon trigger system.

Similarly to electrons, photons are reconstructed from clusters of energy deposits in the electromagnetic calorimeter [21]. After the creation of seed clusters, ID tracks matched to the clusters are used to distinguish electrons, photon conversions and real photons. To discriminate prompt photons from background photons, a photon identification algorithm based on variables characterising the lateral and longitudi-nal shower development in the electromagnetic calorimeter and the shower leakage fraction in the hadronic calorimeter is performed providing *Loose* and *Tight* iden-tification criteria [21, 22]. Photons are required to have $|\eta| < 2.37$ and must be outside the transition region ($1.37 < |\eta| < 1.52$) between barrel and end-caps of the detector [22].

Jets are reconstructed from three-dimensional topological energy clusters of noise-suppressed calorimeter cells [23, 24] using the anti-k_t clustering algorithm [25, 26]. The anti-k_t clustering algorithm is implemented using the FASTJET software package [27]. After the energy calibration [28], which also corrects for additional pile-up contributions to the p_T of each jet according to its area [29, 30], jets are required to have a transverse momentum of $p_T > 20$ GeV. Events containing jets from non-collision sources or detector noise are rejected [31, 32]. To reject jets orig-inating from pile-up interactions, additional requirements are imposed on the track properties of jets with $p_T > 60$ GeV and $|\eta| < 2.4$ [33, 34]. Calibrated $R = 0.4$ jets are reclustered using the anti-k_t clustering algorithm with radius parameters of $R = 1.2$ and $R = 0.8$ to exploit decays of heavy boosted objects (cf. Chap. 11). Using calibrated jets already when running the anti-k_t clustering algorithm with a different radius parameter allows the usage of the jet energy uncertainties estimated for $R = 0.4$ jets [30].

Since hadrons containing b-quarks have a typical decay length of a few millime-tres, it is possible to discriminate between jets originating from b-quarks and jets orig-inating from lighter quarks and gluons within the acceptance of the ID ($|\eta| < 2.5$). The separation is done by combining impact parameter-based and secondary vertex reconstruction algorithms with the topological structure of b- and c-hadron decays inside jets in a multivariate b-tagging discriminant called MV2 [35]. The b-tagging performance is measured by comparing selected $t\bar{t}$ data with MC simulation [36].

Four b-tagging operating points are provided, cutting on the MV2 discriminant for obtaining a 60, 70, 77 or 85% b-tagging efficiency [35]. For the 77% operating point, the c-jet rejection rate is 6, which means the $1/6$ of all c-jets are falsely reconstructed as b-jets, while the light jet rejection rates is estimated to be 134 and the τ-jet rejection rate to 22 [35].

For resolving reconstruction ambiguities, all electrons which have a common ID track with a muon are removed, since the measurement of the MS track clearly indicates that the object was a muon. For resolving reconstruction ambiguities using calorimeter cell information, jets with $p_T > 20\,\text{GeV}$ are removed if they are not b-tagged (using the 85% operating point) and located within $\Delta R_y(\text{jet, electron}) <$ 0.2 of an electron. In case a jet is within $\Delta R_y(\text{jet, muon}) < 0.2$ of a muon and has less than three associated tracks or the ratio between the muon and jet p_T is $p_T(\text{muon})/p_T(\text{jet}) > 0.5$, the jet is also removed since the object is more likely a muon. Subsequently, all leptons within $\Delta R_y(\text{lepton, jet}) < \min\left(\frac{4}{p_T\,[\text{GeV}]}, 0.4\right)$ of a jet are removed in order to reject leptons inside jets arising from decays of c- or b-hadrons.

The negative sum of the transverse momenta of all reconstructed physics objects is called $\mathbf{p}_T^{\text{miss}}$ and its magnitude $|\mathbf{p}_T^{\text{miss}}|$ can be summed with the energy deposits in calorimeter cells not associated with any physics object, the so-called *soft-term* [37], to the missing transverse energy E_T^{miss}. The soft-term is calculated from the ID tracks with $p_T > 400\,\text{MeV}$ originating from the primary vertex but not associated to any physics object in order to account for the jet calibration while maintaining the independence of pile-up [37, 38].

Calculating the missing transverse momentum only considering ID tracks with $p_T > 400\,\text{MeV}$ that are associated with the primary vertex defines the *track-based* missing transverse momentum $E_T^{\text{miss,track}}$. Differences between E_T^{miss} and $E_T^{\text{miss,track}}$ can hint to high pile-up contamination or jet energy mismeasurements where $E_T^{\text{miss,track}}$ is usually smaller than E_T^{miss}.

A more detailed overview of the ATLAS reconstruction software can be found in [39].

References

1. ATLAS Collaboration (2008) The ATLAS experiment at the CERN large hadron collider. JINST 3:S08003. https://doi.org/10.1088/1748-0221/3/08/S08003
2. Wong CY (1995) Introduction to high-energy heavy ion collisions. ISBN: 9789810202637
3. Capeans M et al (2010) ATLAS insertable B-layer technical design report
4. ATLAS Collaboration (2012) ATLAS insertable B-layer technical design report addendum
5. ATLAS Collaboration (2017) Performance of the ATLAS Trigger System in 2015. Eur Phys J C 77:317. https://doi.org/10.1140/epjc/s10052-017-4852-3
6. Abolins M et al (2016) The ATLAS data acquisition and high level trigger system. JINST 11(06):P06008. https://doi.org/10.1088/1748-0221/11/06/P06008
7. Aad G et al (2012) Performance of the ATLAS trigger system in 2010. Eur Phys J C 72:1849. https://doi.org/10.1140/epjc/s10052-011-1849-1. arXiv:1110.1530 [hep-ex]

8. Aad G et al (2013) Technical design report for the phase-I upgrade of the ATLAS TDAQ system
9. ATLAS Collaboration (2011) Luminosity determination in pp collisions at ps = 7 TeV using the ATLAS detector at the LHC. Eur Phys J C 71:1630. https://doi.org/10.1140/epjc/s10052-011-1630-5. arXiv:1101.2185 [hep-ex]
10. van der Meer S (1968) Calibration of the effective beam height in the ISR
11. Rubbia C (1977) Measurement of the luminosity of pp̄ collider with a (generalized) Van Der Meer method
12. ATLAS Collaboration (2016) Luminosity determination in pp collisions at ps = 8 TeV using the ATLAS detector at the LHC. Eur Phys J C 76:653. https://doi.org/10.1140/epjc/s10052-016-4466-1
13. ATLAS Collaboration (2017) Reconstruction of primary vertices at the ATLAS experiment in Run 1 proton-proton collisions at the LHC. Eur Phys J C 77:332. https://doi.org/10.1140/epjc/s10052-017-4887-5
14. ATLAS   Collaboration   (2018).   https://twiki.cern.ch/twiki/bin/view/AtlasPublic/LuminosityPublicResultsRun2
15. Cornelissen T et al (2007) Concepts, design and implementation of the ATLAS new tracking (NEWT)
16. ATLAS Collaboration (2015) Vertex reconstruction performance of the ATLAS detector atps = 13 TeV. ATL-PHYS-PUB-2015-026. https://cds.cern.ch/record/2037717
17. ATLAS Collaboration (2015) Electron identification measurements in ATLAS using ps = 13 TeV data with 50 ns bunch spacing. ATL-PHYS-PUB-2015-041. https://cds.cern.ch/record/2048202
18. Collaboration ATLAS (2017) Electron efficiency measurements with the ATLAS detector using 2012 LHC proton-proton collision data. Eur Phys J C 77:195. https://doi.org/10.1140/epjc/s10052-017-4756-2
19. ATLAS Collaboration (2016) Electron efficiency measurements with the ATLAS detector using the 2015 LHC proton-proton collision data. ATLAS-CONF- 2016-024. https://cds.cern.ch/record/2157687
20. ATLAS Collaboration (2016) Muon reconstruction performance of the ATLAS detector in proton-proton collision data at ps = 13 TeV. Eur Phys J C 76:292
21. ATLAS Collaboration (2016) Measurement of the photon identification efficiencies with the ATLAS detector using LHC Run-1 data. arXiv:1606.01813 [hep-ex]
22. ATLAS Collaboration (2014) Electron and photon energy calibration with the ATLAS detector using LHC Run 1 data. Eur Phys J C 74:3071. https://doi.org/10.1140/epjc/s10052-014-3071-4
23. Lampl W et al (2008) Calorimeter clustering algorithms: description and performance. ATL-LARG-PUB-2008-002. https://cds.cern.ch/record/1099735
24. ATLAS Collaboration (2017) Topological cell clustering in the ATLAS calorimeters and its performance in LHC Run 1. Eur Phys J C 77:490. https://doi.org/10.1140/epjc/s10052-017-5004-5
25. Cacciari M, Salam GP (2006) Dispelling the N3 myth for the kt jetfinder. Phys Lett B 641:57–61. https://doi.org/10.1016/j.physletb.2006.08.037
26. Cacciari M, Salam GP, Soyez G (2008) The anti-k(t) jet clustering algorithm. JHEP 2008(04):063. https://doi.org/10.1088/1126-6708/2008/04/063
27. Cacciari M, Salam GP, Soyez G (2012) FastJet user manual. Eur Phys J C 72:1896. https://doi.org/10.1140/epjc/s10052-012-1896-2
28. ATLAS Collaboration (2015) Jet calibration and systematic uncertainties for jets reconstructed in the ATLAS detector at ps = 13 TeV. ATL-PHYS-PUB-2015-015. https://cds.cern.ch/record/2037613
29. Cacciari M, Salam GP (2008) Pileup subtraction using jet areas. Phys Lett B 659:119–126. https://doi.org/10.1016/j.physletb.2007.09.077
30. ATLAS Collaboration (2017) Jet energy scale measurements and their systematic uncertainties in proton-proton collisions at ps = 13 TeV with the ATLAS detector. Phys Rev D 96:072002. https://doi.org/10.1103/PhysRevD.96.072002

31. ATLAS Collaboration (2015) Selection of jets produced in 13 TeV proton-proton collisions with the ATLAS detector. ATLAS-CONF-2015-029. https://cds.cern.ch/record/2037702
32. ATLAS Collaboration (2010) Data-quality requirements and event cleaning for jets and missing transverse energy reconstruction with the ATLAS detector in proton-proton collisions at a center-of-mass energy of ps = 7 TeV. ATLAS-CONF-2010-038. https://cds.cern.ch/record/1277678
33. ATLAS Collaboration (2014) Tagging and suppression of pileup jets with the ATLAS detector. ATLAS-CONF-2014-018. https://cds.cern.ch/record/1700870
34. ATLAS Collaboration (2016) Performance of pile-up mitigation techniques for jets in pp collisions at ps = 8 TeV using the ATLAS detector. Eur Phys J C 76:581. https://doi.org/10.1140/epjc/s10052-016-4395-z
35. ATLAS Collaboration (2016) Optimisation of the ATLAS b-tagging performance for the 2016 LHC Run. ATL-PHYS-PUB-2016-012. https://cds.cern.ch/record/2160731
36. ATLAS Collaboration (2016) Performance of b-jet identification in the ATLAS experiment. JINST 11:P04008. https://doi.org/10.1088/1748-0221/11/04/P04008
37. ATLAS Collaboration (2015) Performance of missing transverse momentum reconstruction with the ATLAS detector in the first proton-proton collisions at ps = 13 TeV. ATL-PHYS-PUB-2015-027. https://cds.cern.ch/record/2037904
38. ATLAS Collaboration (2015) Expected performance of missing transverse momentum reconstruction for the ATLAS detector at ps = 13 TeV. ATL-PHYSPUB-2015-023. https://cds.cern.ch/record/2037700
39. Aad G et al (2009) Expected performance of the ATLAS experiment - detector, trigger and physics. arXiv:0901.0512 [hep-ex]

Part III
Performance of Muon Reconstruction and Identification

Chapter 7
Muon Reconstruction and Identification

The performance of the physics object identification and reconstruction introduced in Sect. 6.8 needs to be accurately estimated and validated to ensure the success of the ATLAS physics program. Furthermore, with respect to future increases of the instantaneous luminosity, the understanding of the dependence of the particle reconstruction on higher collision and background rates is essential to lead the development of new reconstruction and identification algorithms as well as the underlying detector components. This part of the thesis discusses the current status as well as prospects for the performance of the ATLAS muon reconstruction and identification.

Efficient identification of muon candidates and precise determination of their momenta are key prerequisites for precision measurements with leptonic final states such as the cross section and mass measurements of the W boson [1, 2] or the measurement of the SM Higgs boson coupling properties in decays into four muon final states [3]. The ATLAS Muon Spectrometer permeated by the toroidal magnetic field which allows for the standalone reconstruction of muons is one of the crucial differences in the design of the ATLAS detector with respect to other state-of-the-art high energy particle detectors such as CMS. The reconstruction of muons is complementing the combined reconstruction of muons using measurements from both the Inner Detector and the Muon Spectrometer. This chapter gives a short overview of the muon reconstruction and identification in ATLAS. The measurement of muon reconstruction and track-to-vertex-association efficiencies in LHC Run 2 is discussed in Chap. 8. The behaviour of the Monitored Drift Tube muon chambers under background rates in the range of those expected at the High Luminosity LHC is studied in Chap. 9 exploiting actual pp collision data taken in 2017 with instantaneous luminosities reaching up to $2 \cdot 10^{34} \, \mathrm{cm}^{-2} \, \mathrm{s}^{-1}$.

As described in Sect. 6.4, the MS consists of two types of precision chambers, MDTs and CSCs, and two types of trigger chambers, RPCs and TGCs. The MDTs provide a time and charge measurement which is converted into a drift distance using a time-to-distance relation [4], the CSCs installed in the forward regions ($2.0 < |\eta| < 2.7$) are also providing a time and charge measurement using cathodes segmented by

© Springer Nature Switzerland AG 2019
N. M. Köhler, *Searches for the Supersymmetric Partner of the Top
Quark, Dark Matter and Dark Energy at the ATLAS Experiment*,
Springer Theses, https://doi.org/10.1007/978-3-030-25988-4_7

strips perpendicular and parallel to the wires, but capable of higher counting rates [5] while the RPCs and TGCs provide a position and time measurement where only the position measurement is used in the muon reconstruction.

7.1   Muon Reconstruction

The unique design of the ATLAS muon system embedded in a toroidal magnetic field allows to reconstruct muons without information from the Inner Detector (ID). In order to reconstruct a muon in the MS, so-called muon segments (short straight-line tracks) are identified from which a track-finding algorithm builds muon tracks. Segments are found in the MDT (CSC) chambers based on certain hit (strip) requirements and angular restrictions from the position measurements performed by the RPC and TGC chambers [6]. The track-finding algorithm ranks the segments by their quality and starts to find tracks based on the high quality segments of the middle chamber layers. Then, the algorithm extends the search to the outer layers and finally also to the inner ones. First, different segments are matched by their position and angle. At least two matched segments are required to build a track in all regions but the transition region between barrel and endcap. There, one segment with high quality can be sufficient. The hits belonging to a segment along the track trajectory are fitted using a global χ^2 fit [7]. In case the χ^2 fit converges and satisfies certain selection criteria, the track is accepted and the associated segments cannot be used as a seed for new track candidates anymore.

By using the fitted track parameters and the geometrical description of the muon chambers, the expected signals in any tube, wire or strip along the track are computed. In case a hit is expected but not found in a particular tube, wire or strip, a so-called *hole* is defined. If a measured hit on the track deviates from its predicted position by more than 2.5 standard deviations, the hit is removed from the track fit and flagged as an *outlier* [8].

If more than 10% of a track's hits are associated to another track, the track with the worse track quality—defined in terms of the χ^2 of the fit as well as the number of hits and holes of the track—is rejected. In order to ensure high efficiency for close-by muons, no rejection is made for tracks passing three chamber layers but sharing no hits in one of them. All remaining tracks are saved as so-called muon spectrometer tracks (MS tracks). If an MS track is successfully extrapolated to the interaction point and refitted taking into account the energy loss in the calorimeter, a hit recovery and hole search is performed including all chamber layers of the MS. After resolving the reconstruction ambiguities due to overlapping hits as done for the MS tracks, the remaining tracks are saved as so-called muon extrapolated tracks (ME tracks).

ATLAS Muon Types

Five complementary types of muon reconstruction algorithms are used within the ATLAS reconstruction software.

- For the region outside the ID coverage ($2.5 < |\eta| < 2.7$), muons are reconstructed by the MS only. Thus, **ME tracks** are used as muons for this case.
- Muons reconstructed by looking for an ID track in a cone around an ME track and calculating a matching χ^2 using the track parameters of the tracks defined at the interaction point and the entrance to the MS, are referred to as **combined muons** since the information of both the ID and MS is combined in order to minimise the uncertainties on the track parameters. For each combined muon, the hit recovery and hole search are executed and additional requirements on the momentum compatibility between the measurements of ID and MS as well as the matching χ^2 and the magnetic field properties are made. Finally, all MS hits of the combined muon are re-calibrated using the parameters from the combined fit of the hits in the ID and MS in order to obtain the best possible momentum resolution. Since the reconstruction of combined muons relies on the presence of an ME track, the algorithm becomes inefficient for regions with poor MS coverage.
- In contrary to combined muons, muons can also be reconstructed based on ID tracks with $p_T > 2\,\text{GeV}$ which are extrapolated to the MS where a segment search based on all MS hits found along the extrapolated trajectory is initiated. The reconstructed segments are matched to the ID track using a neural network based on the segment and ID track parameters. If at least one segment is matched to an ID track, the same fit and selection procedure as for combined muons is performed resulting in so-called **MuGirl muons** [8]. However, the main disadvantage of MuGirl muons with respect to combined muons is the much higher combinatorics when starting the reconstruction based on ID tracks which results in a tremendously larger computational effort. Furthermore, since the ID tracks are matched to segments in the MS, no independent track finding was performed in the MS which increases the probability to misidentify secondary muons arising from pion or kaon decays as muons originating from the hard process.
- In case an ID track with $p_T > 2\,\text{GeV}$ extrapolated to the MS can be matched to an MDT or a CSC muon segment, a so-called **segment-tagged muon** is defined. Thereby, it is allowed that the innermost chamber layer does not contain a matched segment since a muon with low transverse momenta can undergo multiple scatterings and putting requirements on the number of holes to find segments could lead to a drop of reconstruction efficiency. In order to reduce the contribution of misidentified secondary muons arising from pion or kaon decays muons, additional requirements are made on the quality of the segments [8].
- In regions with poor coverage by the MS, in especially $|\eta| < 0.1$, where the MS is only partially instrumented to allow for cabling and services to the calorimeters and the ID, isolated ID tracks with $p_T > 5\,\text{GeV}$ are extrapolated through the calorimeters. In case the total energy deposit of the ID track in the electromagnetic and hadronic calorimeters is compatible to the hypothesis of a minimum-ionising particle, a so-called **calorimeter-tagged muon** is identified [9].

The different types of muon reconstruction algorithms are visualised in Fig. 7.1.

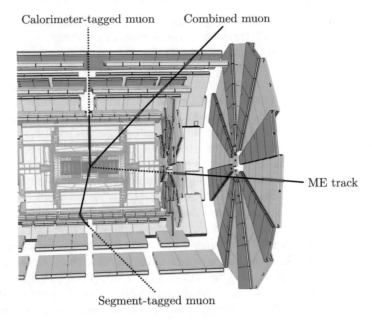

Fig. 7.1 Schematic drawing visualising the muon reconstruction algorithms exploited by ATLAS

7.2   Muon Identification

The presence of secondary muons arising from pion or kaon decays, which are also referred to as *fake* muons, needs to be suppressed by applying additional muon identification criteria. A range of different identification criteria are defined to meet varying requirements on the selection efficiency and robustness of the momentum measurement depending on the use case.

For all muon types using information from the ID, the following series of quality requirements is applied in order suppress muons originating from in-flight decays of charged hadrons in the ID which can be recognised through kinks in the reconstructed track [10].

- The ID track needs to have a least one hit in the pixel detector.
- The ID track needs to have a least five hits in the SCT detector.
- The ID track needs to contain less than 3 holes in the pixel and the SCT detector, respectively.

A missing hit in the ID is only considered as a hole in case it is located between hits which were successfully associated to the ID track.

To estimate the quality of a fitted combined muon track, in addition to the normalised χ^2 of the track fit, the significance of the absolute difference between the charge-over-momentum measurements performed in the ID and the MS,

$$\left(\frac{q}{p}\right)^{\text{sig}} = \frac{\left|\left(\frac{q}{p}\right)_{\text{ID}} - \left(\frac{q}{p}\right)_{\text{ME}}\right|}{\sqrt{\sigma_{\text{ID}}^2 + \sigma_{\text{ME}}^2}} \quad , \tag{7.1}$$

was found to be an effective discriminating variable [10]. Additionally, the ratio of the absolute difference between the transverse momenta measured in the ID and the MS and the transverse momentum p_{T}^{CB} of the combined track,

$$\rho' = \frac{p_{\text{T}}^{\text{ID}} - p_{\text{T}}^{\text{ME}}}{p_{\text{T}}^{\text{CB}}} \quad , \tag{7.2}$$

provides a good suppression against fake muons.

Four muon identification criteria [10] are defined:

- The *Medium* identification criteria is the default in ATLAS. It is designed to provide a balance between high efficiency and purity while keeping systematic uncertainties associated with the muon reconstruction and calibration minimal. Only ME tracks in $2.5 < |\eta| < 2.7$ which have hits in at least three MDT/CSC layers and combined muons with at least three hits on at least two MDT layers for $0.1 < |\eta| < 2.5$ are used. For the region $|\eta| < 0.1$, combined muons with tracks in at least one MDT layer but no more than one hole in an MDT layer are required. In addition the requirement $\left(\frac{q}{p}\right)^{\text{sig}} < 7$ suppresses contributions from hadrons being mis-identified as muons. The MuGirl algorithm recovers the small fraction of events where the combined algorithm was inefficient due to poor MS coverage.
- The *Loose* identification criterion is designed to maximise the reconstruction efficiency while ensuring that the quality of the muon tracks is kept at a sufficient level. It uses all muons passing the *Medium* identification criterion and in addition calorimeter-tagged and segment-tagged muons are used for $|\eta| < 0.1$ where the reconstruction efficiency of combined muons is low due to the partial instrumentation of the MS. ME track muons are used for $2.5 < |\eta| < 2.7$, where no ID tracks are available.
- The *Tight* identification criterion is designed to minimise the number of fake muons at the price of some loss in reconstruction efficiency and larger systematic uncertainties. Only combined muons satisfying the *Medium* identification criteria and having hits in at least two chamber layers are considered. Furthermore, a requirement on the normalised χ^2 ($\chi^2/N_{\text{DOF}} < 8$) as well as p_{T}-dependent requirements on $\left(\frac{q}{p}\right)^{\text{sig}}$ and ρ' are chosen such that the efficiency for selecting prompt muons is above 95% (90%) for $p_{\text{T}} > 9\,\text{GeV}$ ($p_{\text{T}} > 4\,\text{GeV}$) while the rejection of fake muons is maximised.
- The *High-p_T* identification criteria provides the best possible momentum resolution for muons with transverse momenta above $100\,\text{GeV}$. Only combined muons passing the *Medium* identification criteria and having at least three hits in three different MDT or CSC layers are considered. Additionally, regions where the alignment of the chambers is not fully validated yet are vetoed as a precaution.

References

1. ATLAS Collaboration (2018) Measurement of the W-boson mass in pp collisions at $\sqrt{s} =$ 7 TeV with the ATLAS detector. Eur Phys J C 78:110. https://doi.org/10.1140/epjc/s10052-017-5475-4
2. ATLAS Collaboration (2016) Measurement of $W^{\pm}$ and Z-boson production cross sections in pp collisions at $\sqrt{s} = 13$ TeV with the ATLAS detector. Phys Lett B 759:601. https://doi.org/10.1016/j.physletb.2016.06.023
3. ATLAS Collaboration (2017) Measurement of the Higgs boson coupling properties in the $H \rightarrow ZZ^{*} \rightarrow 4\ell$ decay channel at $\sqrt{s} = 13$ TeV with the ATLAS detector. arXiv:1712.02304 [hep-ex]
4. Bagnaia P et al (2008) Calibration model for the MDT chambers of the ATLAS muon spectrometer
5. Nikolopoulos K et al (2008) Cathode strip chambers in ATLAS: installation, commissioning and in situ performance. In: Proceedings, 2008 IEEE nuclear science symposium, medical imaging conference and 16th international workshop on room-temperature semiconductor X-ray and gamma-ray detectors (NSS/MIC 2008/RTSD 2008), Dresden, Germany, 19–25 October 2008, pp 2819–2824. https://doi.org/10.1109/NSSMIC.2008.4774958
6. van Eldik N (2007) The ATLAS muon spectrometer. PhD thesis, University of Massachusetts, Amherst. http://weblib.cern.ch/abstract?CERN-THESIS-2007-045
7. Cornelissen TG et al (2006) Track fitting in the ATLAS experiment. http://cds.cern.ch/record/1005181. Accessed on 12 Dec 2006
8. van Kesteren Z (2010) Identification of muons in ATLAS. PhD thesis, University of Amsterdam. http://www.nikhef.nl/pub/services/biblio/theses_pdf/thesis_Z_v_Kesteren.pdf
9. Ordonez Sanz G (2009) Muon identification in the ATLAS calorimeters. https://cds.cern.ch/record/1196071. Accessed on 12 Jun 2009
10. ATLAS Collaboration (2016) Muon reconstruction performance of the ATLAS detector in proton-proton collision data at $\sqrt{s} = 13$ TeV. Eur Phys J C 76:292

Chapter 8
Muon Efficiency Measurement

A precise understanding of the muon reconstruction and identification efficiency becomes of increasing importance the more muons are present in the final state under consideration since the event yield scales with the efficiency to the power of the number of muons to be identified. For muons with $|\eta| < 2.5$, the reconstruction and identification efficiency can be measured in collision data using di-muon decays of SM particles such as the Z boson or the J/ψ meson by means of the so-called *tag-and-probe* method. Section 8.1 introduces the tag-and-probe methodology on the example of $Z \rightarrow \mu^+\mu^-$ decays. Section 8.2 presents the measurement of the reconstruction and identification efficiencies for muons with $p_T > 10\,\text{GeV}$ performed in LHC Run 2. The measurement of the selection efficiency for the track-to-vertex association of muons is discussed in Sect. 8.3.

8.1 Tag-and-Probe Methodology

By requiring one well-reconstructed, isolated muon and an additional isolated ID or ME track of opposite charge with the two of them having an invariant mass compatible with the one of the Z boson, one likely has reconstructed a $Z \rightarrow \mu^+\mu^-$ decay. The well-reconstructed muon which also caused a single muon trigger to record the event is commonly referred to as a *tag* since it is the basic requirement that the event is considered in the measurement. The isolated ID or ME track is referred to as a *probe* since due to the requirements on its charge and the invariant mass, it is very likely to be the second muon from the $Z \rightarrow \mu^+\mu^-$ decay and it can be probed if it was indeed reconstructed as such. Figure 8.1 shows a schematic view of the ATLAS detector with the tag muon (marked in green) and an ID or ME probe (marked in red) with an invariant mass compatible with the one of a Z boson. The ID or ME probe is not necessarily identified as a muon by the identification criteria described in Sect. 7.2.

© Springer Nature Switzerland AG 2019
N. M. Köhler, *Searches for the Supersymmetric Partner of the Top Quark, Dark Matter and Dark Energy at the ATLAS Experiment*, Springer Theses, https://doi.org/10.1007/978-3-030-25988-4_8

Fig. 8.1 Schematic drawing of the ATLAS detector systems involved in the muon reconstruction as well as a tag-and-probe pair which is potentially arising from a Z boson decay. The tag muon (green) is a well-reconstructed and isolated muon, the probe (red) can be an ID or ME track

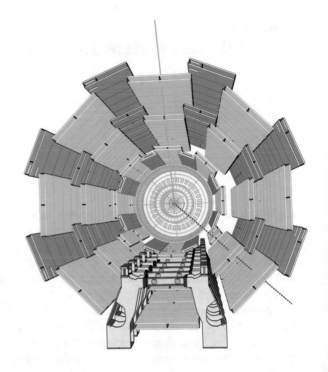

8.2 Reconstruction Efficiency

As track reconstruction is performed independently in the ID and the MS, the total efficiency of a muon to be reconstructed and identified by an identification criterion X is given by the product of the efficiency that the muon is identified in the ID, and the efficiency that, given it was successfully identified in the ID, it is also identified as a muon passing criterion X. Assuming that the ID track reconstruction efficiency is independent of the MS track reconstruction ($\epsilon(\text{ID}) = \epsilon(\text{ID}|\text{ME})$), the former efficiency can be measured by using ME tracks as probes and testing whether a reconstructed ID tracks is found within $\Delta R < 0.05$ of the ME track, $\epsilon(\text{ID}|\text{ME})$. The latter can be measured by using a calorimeter-tagged (CT) muon[1] and testing whether a muon was reconstructed and identified using the identification criterion X, $\epsilon(\text{X}|\text{CT})$ within $\Delta R < 0.05$ of the CT probe. The occurrence of a reconstructed ID track or muon, respectively, within $\Delta R < 0.05$ of the corresponding probe is referred to as a *match* and the total efficiency is given by

$$\epsilon(\text{X}) = \epsilon(\text{X}|\text{CT}) \cdot \epsilon(\text{ID}|\text{ME}) = \frac{N_{\text{X-match | CT probe}}}{N_{\text{CT probe}}} \cdot \frac{N_{\text{ID track-match | ME probe}}}{N_{\text{ME probe}}} \quad (8.1)$$

---

[1]Since using ID tracks as probes is more prone to be contaminated by fake muons and resulting in a much larger combinatorial factor, calorimeter-tagged muons are used as probes, also being fully unbiased with respect to the muon reconstruction in the MS.

can be directly evaluated for the *Medium*, *Tight* and *High-p_T* identification criteria. For the *Loose* criteria however, it can only be evaluated for muons with $|\eta| > 0.1$ in this way since calorimeter-tagged muons are part of the *Loose* identification criteria for $|\eta| < 0.1$. In this region, the efficiency of calorimeter-tagged muons is measured separately using ME tracks as probes.

The tag muon is required to satisfy the *Medium* identification criteria, $p_T > 28$ GeV, $|\eta| < 2.5$ and the single muon trigger requirement described in Table A.1 in the Appendix. In order to suppress secondaries from jet activity, a selection on the ratios $p_T^{\text{varcone}30}/p_T$ and $E_T^{\text{cone}20}/E_T$ (as introduced in Sect. 11.2), designed to be 99% efficient independent of the muon p_T is applied as isolation criterion.

The invariant mass of the tag-and-probe pair is required to be $|m_{\text{tag-probe}} - m_Z| < 10$ GeV. Furthermore, the probes also have to satisfy $p_T > 10$ GeV, $|\eta| < 2.5$ and the same isolation criteria as the tag muon. CT probes are additionally required to satisfy the ID track quality requirements introduced in Sect. 7.2. The selection criteria applied for the different tag-and-probe selections are summarised in Table A.1 in the Appendix.

Figure 8.2 shows the CT probe distributions of η and p_T after applying the selection criteria summarised Table A.1 on both recorded and simulated data. The SM processes of $Z(\to \mu^+\mu^-)$+jets, diboson, $Z(\to \tau^+\tau^-)$+jets and $W(\to \mu\nu)$+jets production are simulated using the POWHEG- BOX 2 [1] generator at NLO with the CT10 PDF set [2] for the hard scatter process. For the parton showering PYTHIA 8.186 [3] with the CTEQ6L1 [4] PDF set and the AZNLO [5] underlying event tune are used. SM $t\bar{t}$ production is generated using POWHEG- BOX 2 generator at NLO with the NNPDF3.0NLO PDF set [6] while the parton showering is performed in PYTHIA 8.186 with the NNPDF2.3LO PDF set [7] and A14 [8] as the underlying event tune. The simulation of multi-jet events involving heavy-flavour jets, namely $b\bar{b}$ and $c\bar{c}$ production, is performed using PYTHIA8B [3] using the NNPDF2.3LO PDF set and A14 as the underlying event tune. For all simulations but the multi-jet contributions, the hadronisation of b- and c-hadron decays is performed with EVT-GEN 1.2.0 [9]. The matrix element generators, parton showering prescriptions, parton distribution function sets, underlying-event tune parameters and cross-section orders for all simulated processes are summarised in Table A.2 in the Appendix.

After applying the tag-and-probe selection, the purity of $Z \to \mu^+\mu^-$ decays in the probe distributions is about 99.9%. Diboson production involving $Z \to \mu^+\mu^-$ decays, irreducible contributions from $Z(\to \tau^+\tau^-)$+jets and $t\bar{t}$ as well as reducible contributions arising from $W(\to \mu\nu)$+jets and multi-jet events where probes are mainly originating from secondary muons from pion, kaon or heavy-flavour decays comprise the remaining 0.1%. As depicted in Fig. 8.2, the multi-jet and $W(\to \mu\nu)$+jets contributions are seen to be poorly described by the MC simulation.

Background Estimation

Diboson production involving $Z \to \mu^+\mu^-$ decays is treated as part of the signal processes for this measurement. The normalisation of the irreducible background contributions from $Z(\to \tau^+\tau^-)$+jets and $t\bar{t}$ production is estimated from MC simulation.

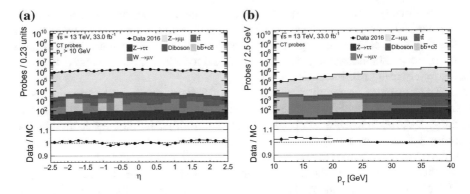

Fig. 8.2 Comparison of pp collision data and MC prediction for the distributions of η (**a**) and p_T (**b**) selecting calorimeter-tagged probes. The MC predictions were scaled to an integrated luminosity of $33.0\,\mathrm{fb^{-1}}$ corresponding to the dataset collected in 2016. No data-driven SM background estimation is applied

Since the MC prediction reducible backgrounds is statistically very limited, a data-driven estimate is used. For such backgrounds, the charge of tag muon and the probe are uncorrelated, since the two objects are not arising from a Z boson or $t\bar{t}$ decay. Thus, the normalisation of the background contribution is estimated from data using events for which the tag and probe have the same charge (SC), but pass all other requirements of the selection. To account for possible residual correlations between the tag and probe charge in the background, the ratio between the background yield in an anti-isolated OC region and the background yield in the corresponding SC region is calculated and multiplied as a transfer factor to the SC estimate. This method (visualised in Fig. 8.3) is also referred to as *ABCD-method*. For the anti-isolated regions, the isolation requirement on E_T^{cone20}/E_T is changed to $E_T^{cone20}/E_T > 0.2$ suppressing isolated probes and the invariant mass window is widened to $|m_{\text{tag}-\text{probe}} - m_Z| < 20\,\mathrm{GeV}$ in order to enhance the statistics for the estimation of the transfer factor. Before calculating the transfer factor on data in the anti-isolated regions, the signal and irreducible background contributions estimated from MC simulation, $N_{\text{sig}}^{\text{MC,anti-iso}}$ and $N_{\text{irr}}^{\text{MC,anti-iso}}$, respectively, are subtracted, resulting in a transfer factor

$$T_{\text{OC/SC}} = \frac{N^{\text{Data,anti-iso}}(\text{OC}) - N_{\text{sig}}^{\text{MC,anti-iso}}(\text{OC}) - N_{\text{irr}}^{\text{MC,anti-iso}}(\text{OC})}{N^{\text{Data,anti-iso}}(\text{SC}) - N_{\text{sig}}^{\text{MC,anti-iso}}(\text{SC}) - N_{\text{irr}}^{\text{MC,anti-iso}}(\text{SC})}. \tag{8.2}$$

The expected reducible background contribution in the probe distribution $N^{\text{red,iso}}$ (OC), it follows

$$N^{\text{red,iso}}(\text{OC}) = T_{\text{OC/SC}} \cdot \left(N^{\text{Data,iso}}(\text{SC}) - N_{\text{sig}}^{\text{MC,iso}}(\text{SC}) - N_{\text{irr}}^{\text{MC,iso}}(\text{SC}) \right), \tag{8.3}$$

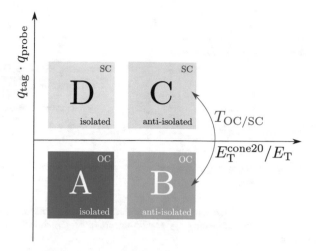

Fig. 8.3 Illustration of the ABCD method applied for the data-driven estimation of reducible backgrounds in the measurement of muon reconstruction efficiencies

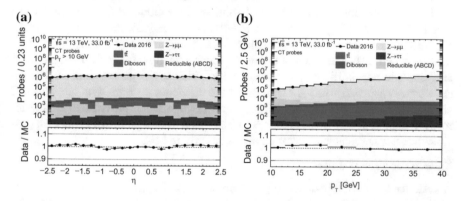

Fig. 8.4 Comparison of pp collision data and MC prediction for the distributions of η (**a**) and p_T (**b**) selecting calorimeter-tagged probes after the reducible background contributions where estimated using the ABCD method. The MC predictions were scaled to an integrated luminosity of $33.0\,\text{fb}^{-1}$ corresponding to the dataset collected in 2016

where also for the SC region, the contributions from the signal and the irreducible backgrounds estimated from MC are subtracted. The shape of the reducible background contribution is taken from the SC region but its normalisation from the transfer factor. Figure 8.4 depicts the pseudorapidity and the transverse momentum of the calorimeter-tagged probes compared to predictions obtained using the data-driven background subtraction. Compared to Fig. 8.2, the data-driven reducible background estimate is not limited by statistics anymore as it was when taken from MC and the agreement of MC prediction and data is clearly improved in the low transverse momenta regime of the probe distribution.

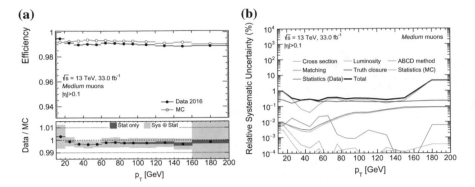

Fig. 8.5 Muon reconstruction efficiencies of the *Medium* identification criteria measured in data and MC (**a**) as well as the corresponding systematic uncertainties (**b**) shown against the muon transverse momentum

In order to calculate the reconstruction efficiency using collision data, the irreducible and reducible background contributions are subtracted from the probe and match distributions, respectively,[2] and the efficiencies are calculated using Eq. (8.1). Figure 8.5(a) shows the muon reconstruction efficiency for the *Medium* muon identification criterion as a function of the muon transverse momentum for both recorded and simulated data. The muon reconstruction efficiency is greater than 98.5% and shows no dependency on p_T. The ratio of the efficiency measured in data and MC simulation is referred to as the *efficiency scale factor* which is provided to all physics measurements and searches in ATLAS to be applied to simulation to account for possible detector inefficiencies not modelled correctly in MC.

Systematic Uncertainties

A systematic uncertainty on the application of the ABCD method is defined by also estimating the transfer factor from simulated $W(\rightarrow \mu\nu)$+jets and multi-jet events and taking the envelope of the data-driven, the simulated and a transfer factor of 1 assuming a perfect OC/SC symmetry of the reducible background. Since the statistics of the SC distribution in data decreases with increasing transverse momenta, a small fluctuation in data can lead to a significant envelope applying the varied transfer factors (cf. Fig. 8.5b). The transfer factor obtained from calorimeter-tagged muons is evaluated to be $T_{OC/SC} = 1.507 \pm 0.005(\text{stat})^{+0.219}_{-0.507}(\text{sys})$ while the one from ME tracks is $T_{OC/SC} = 1.895 \pm 0.005(\text{stat})^{+0.0}_{-0.895}(\text{sys})$.

Since the matching between probe and match is performed by a $\Delta R < 0.05$ requirement, a systematic uncertainty is reflecting the arbitrariness in this choice is obtained by varying the requirement on ΔR by a factor of two and one-half, respectively. The resulting uncertainty (shown in blue in Fig. 8.5b) is negligible compared to the one on the application of the ABCD method. Also the variation of the integrated

---

[2]The transfer factor is evaluated from the anti-isolated probe distributions in both cases.

luminosity by its uncertainty (bright green) as well as varying the cross section of the $Z(\rightarrow \mu^+\mu^-)$+jets simulation by its uncertainty (grey) are of negligible impact.

The reconstruction algorithm of calorimeter-tagged muons rejects muons with a high energy loss. Since these are also less likely to pass the combined muon reconstruction, a slight upward bias of the efficiency measured using calorimeter-tagged muon probes instead of ID track probes is observed. Studies at generator-level confirm the efficiencies measured with ID track probes and a systematic uncertainty is assigned to the measurement to account for this. This source of uncertainty is commonly referred to as *Truth closure* [10]. Since the effect is expected to have the same impact on measured and simulated data and thus cancel out in the efficiency scale factor, a conservative uncertainty of half the envelope of MC efficiency and the efficiency at generator-level is assigned to the efficiency scale factor.

At very high transverse momenta above 200 GeV, the efficiency slightly degrades since a very high energy loss in the calorimeter can affect the efficiency of the muon reconstruction algorithm. Thus, for muons with $p_T > 200$ GeV, a p_T-dependent systematic uncertainty is assigned calculated by measuring MC efficiencies using high-mass Drell-Yan MC and linearly interpolating the MC efficiencies for $p_T > 200$ GeV. This approach is conservative enough to cover the case where the degradation of reconstruction efficiency due to very high energy loss in the calorimeter can be twice as large in data than in MC.

However, Fig. 8.5(a) indicates shortcomings of the application of the ABCD method at low and very high probe p_T. This is due to limited statistics in the auxiliary regions used to perform the estimate. In addition, the assumption of the method that isolation and charge are not correlated does not fully hold. This is mainly resolved by performing a dedicated tag-and-probe analysis exploiting $J/\psi \rightarrow \mu^+\mu^-$ decays for transverse momenta below 15 GeV (cf. Fig. 8.6). For muons with large transverse momenta, the contribution of reducible backgrounds strongly decreases resulting as much less populated SC regions needed for the data-driven estimate. This leads to unnaturally large systematic uncertainties.

Reducible Background Estimation Using a Template Fit

In order to resolve the shortcomings of the ABCD-method described above, a simultaneous fit is introduced, removing the necessity of the estimation of a OC/SC transfer factor in anti-isolated regions. In contrary to the ABCD-method, all SM processes producing an opposite-charge muon pair are considered as signal which only leaves reducible background processes. In simulation, both the tag muon and the probe are required to be a muon at generator-level in order to eliminate any background contaminating the simulated sample. For the estimation of the reducible background contributions in data, the whole SC distribution is assumed to arise from non-prompt muons.

The invariant mass spectrum of the tag-and-probe pairs in the SC region is expected to have a smooth monotone shape since no resonant decay is present. The shape can be approximately parametrised using the functional form

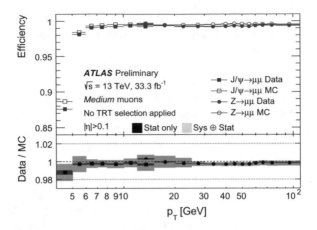

Fig. 8.6 Muon reconstruction efficiencies for the Medium identification algorithm measured in $J/\psi \rightarrow \mu^+\mu^-$ and $Z \rightarrow \mu^+\mu^-$ events as a function of the muon momentum. The measurement is based on 33.3 fb^{-1} of pp collision data taken in 2016. The prediction by the detector simulation is depicted as empty circles (squares), while the full circles (squares) indicate the observation in collision data for $Z \rightarrow \mu^+\mu^-$ ($J/\psi \rightarrow \mu^+\mu^-$) events. Only statistical errors are shown in the top panel. The bottom panel reports the efficiency scale factors. The darker error bands indicate the statistical uncertainty, while the lighter bands indicate the quadratic sum of statistical and systematic uncertainties [11]

$$f\left(m_{\text{tag}-\text{probe}}\right) = \left(1 - \frac{m_{\text{tag}-\text{probe}}}{\Lambda}\right)^{p_1} \cdot \left(\frac{m_{\text{tag}-\text{probe}}}{\Lambda}\right)^{p_2} \tag{8.4}$$

where p_1 and p_2 are free parameters and Λ is the energy necessary to produce the di-muon pair [12]. Since the fit needs to be performed separately for each bin of the distribution of interest, the invariant mass requirement is widened to 61 GeV < $m_{\text{tag}-\text{probe}}$ < 117 GeV and Λ is chosen to be twice the upper border of the invariant mass window. A simultaneous fit is performed using Eq. (8.4) for fitting the shape of the SC distribution in data (cf. Fig. 8.7a) and the weighted sum of the probability density functions obtained from the SC fit and the shape of the OC MC simulation for fitting the OC distribution in data (cf. Fig. 8.7b). In case the SC distribution in data contains less than three data events, a constant is fitted as the shape of the contribution of non-prompt muons in the OC fit.

Figure 8.8(a) shows the reconstruction efficiencies for the *Medium* identification criterion for both data and simulation applying the simultaneous fit in order to estimate the reducible background contributions. Compared to Fig. 8.5, also the efficiency measured in data is almost completely flat in p_T and a constant efficiency scale factor is obtained. The systematic uncertainties (cf. Fig. 8.8b) are significantly reduced and dominated by the *Truth closure* component discussed above. Two types of systematic uncertainties are introduced due to the application of the simultaneous fit. In order to account for the uncertainty due to fitting the shape of the SC distribution, the SC fit parameters p_1 and p_2 are varied by their uncertainty when performing the simultaneous fit resulting in the *Template shape* uncertainty. The uncertainty on

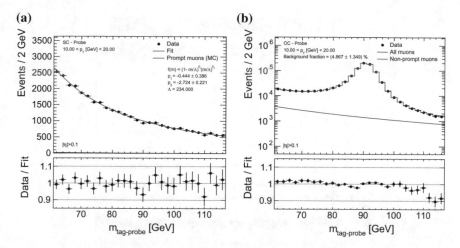

Fig. 8.7 Simultaneous fit performed on SC data (**a**) using Eq. (8.4) and OC data (**b**) using the shape of the SC region as the background and the shape of the signal MC simulation as the signal component of the fit, respectively. Calorimeter-tagged probes with $10\,\text{GeV} < p_T < 20\,\text{GeV}$ and $|\eta| > 0.1$ are shown. In **a**, the negligible fraction of simulated signal events contaminating the SC region is drawn in red

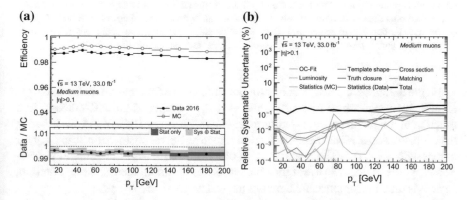

Fig. 8.8 Muon reconstruction efficiencies of the *Medium* identification criteria measured in data and MC (**a**) as well as the corresponding systematic uncertainties (**b**) applying a simultaneous fit in order to estimate the contribution of non-prompt muons

the fitted background fraction in the OC region is estimated by varying the fitted OC/SC fraction by its uncertainty (labelled as OC-*Fit* in Fig. 8.8b). Furthermore, the uncertainty on the $\Delta R < 0.05$ matching requirement between probes and matches is reduced by comparing the ΔR matching to the actual physics objects information inside the analysis software environment and taking the envelope of both procedures as an uncertainty.

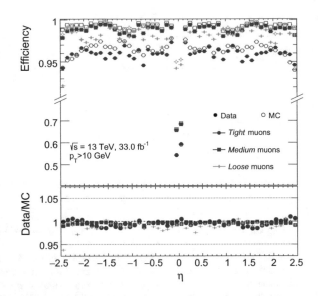

Fig. 8.9 Muon reconstruction efficiencies for the *Loose/Medium/Tight* identification algorithms measured in $Z \rightarrow \mu^+\mu^-$ events of the 2016 dataset as a function of the muon pseudorapidity for muons with $p_T > 10$ GeV. The prediction by the detector simulation is depicted as open circles, while filled dots indicate the observation in collision data with statistical errors. The bottom panel shows the ratio between expected and observed efficiencies, the efficiency scale factor. The errors in the bottom panel show the quadratic sum of statistical and systematic uncertainty

Figure 8.9 shows the muon reconstruction efficiencies for the *Loose/Medium/Tight* identification algorithms measured in $Z \rightarrow \mu^+\mu^-$ events of the 2016 dataset as a function of the muon pseudorapidity for muons with $p_T > 10$ GeV. As described in Sect. 7.2, the *Loose* identification criterion is recovering reconstruction efficiency for muons with $|\eta| < 0.1$. The reconstruction efficiency of the *Tight* identification criterion is approximately 2% lower than the one for the *Medium* identification criterion, since it is tuned to suppress fake muons as well as possible.

Figure 8.10 shows the efficiency scale factor for the *Medium* identification criterion in the η-ϕ plane of the detector. The binning reflects the geometry of the ATLAS muon spectrometer. The agreement between reconstruction efficiencies measured in data and simulation is between 95 and 101% throughout the whole coverage of the *Medium* identification criterion. The reconstruction efficiencies measured on data and MC as well as the efficiency scale factor for all muon identification criteria can be found in Appendix A.1.1.

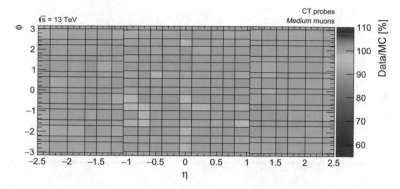

Fig. 8.10 Muon reconstruction efficiency scale factors for the *Medium* identification algorithm measured in $Z \to \mu^+ \mu^-$ events of the 2016 dataset as a function of the muon η and ϕ for muons with $p_T > 10$ GeV. The binning reflects the geometry of the ATLAS muon spectrometer

8.3 Track-to-Vertex Association Efficiency

As described in Sect. 11.2, muons used in precision measurements or searches are often restricted to satisfy $|z_0| \cdot \sin(\theta) < 0.5$ mm and $|d_0|/\sigma_{d_0} < 3$ in order to ensure that they are originating from the primary vertex. These selection criteria are also referred to as *track-to-vertex association* requirements. Since they are not part of the muon identification criteria, their selection efficiency has to be estimated independently in addition to the muon reconstruction and identification efficiency. The measurement of the track-to-vertex association efficiency of muons with low transverse momenta by means of $J/\psi \to \mu^+ \mu^-$ decays is not feasible since due to the lifetime of the J/ψ meson, the number of tag-and-probe pairs passing the track-to-vertex association requirements is strongly suppressed. With the introduction of the simultaneous fit described above, the track-to-vertex association efficiency can be measured in $Z \to \mu^+ \mu^-$ events with high precision also for low transverse momenta and a good suppression of reducible backgrounds.

The tag selection is identical to the one from the muon reconstruction efficiency measurement summarised in Table A.1 in the Appendix. Muons satisfying the *Loose* identification criterion as well as $|\eta| < 2.5$ and $p_T > 10$ GeV are used as probes. The selection efficiency of requiring $|z_0| \cdot \sin(\theta) < 0.5$ mm and $|d_0|/\sigma_{d_0} < 3$ is shown in Fig. 8.11 depending on the muon p_T and the average number of interactions per bunch crossing $<\mu>$. The track-to-vertex association efficiency is depending on the successful reconstruction of the primary vertex as well as the precision of the measurement of the impact parameters d_0 and z_0. It is more likely that the primary vertex is reconstructed correctly in case of the presence of high-p_T objects in the event. Furthermore, tracks of high-p_T muons are less bended and therefore result in a more precise measurement of the impact parameters. Both effect result in an increasing track-to-vertex association efficiency with the transverse momentum of

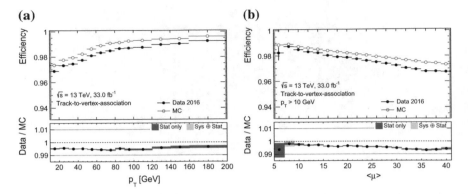

Fig. 8.11 Muon track-to-vertex association efficiency measured in $Z \to \mu^+\mu^-$ events of the 2016 dataset as a function of the muon transverse momentum (**a**) and the average number of interactions per bunch crossing (**b**). The bottom panel shows the ratio between expected and observed efficiencies, the efficiency scale factor

the muon.[3] The average number of interactions per bunch crossing is proportional to the instantaneous luminosity. Thus, for large instantaneous luminosities, an increased number of vertices is expected resulting in a decreasing track-to-vertex association efficiency (cf. Fig. 8.11b).

References

1. Alioli S et al (2010) A general framework for implementing NLO calculations in shower Monte Carlo programs: the POWHEG BOX. JHEP 2010(6):043. https://doi.org/10.1007/JHEP06(2010)043. arXiv:1002.2581 [hep-ph]
2. Lai H-L et al (2010) New parton distributions for collider physics. Phys Rev D 82:074024. https://doi.org/10.1103/PhysRevD.82.074024
3. Sjöstrand T, Mrenna S, Skands PZ (2008) A brief introduction to PYTHIA 8.1. Comput Phys Commun 178:852–867.https://doi.org/10.1016/j.cpc.2008.01.036
4. Pumplin J et al (2002) New generation of parton distributions with uncertainties from global QCD analysis. JHEP (2002)07:012. https://doi.org/10.1088/1126-6708/2002/07/012
5. ATLAS Collaboration (2014) Measurement of the Z/γ^* boson transverse momentum distribution in pp collisions at ps = 7 TeV with the ATLAS detector. JHEP 2014(09):145. https://doi.org/10.1007/JHEP09(2014)145
6. Ball RD et al (2015) Parton distributions for the LHC Run II. JHEP 2015(04):040. https://doi.org/10.1007/JHEP04(2015)040
7. Ball RD et al (2013) Parton distributions with LHC data. Nucl Phys B 867:244–289. https://doi.org/10.1016/j.nuclphysb.2012.10.003
8. ATLAS Collaboration (2014) ATLAS Pythia 8 tunes to 7 TeV data. ATL-PHYSPUB- 2014-021. https://cds.cern.ch/record/1966419

---

[3]Here, muons satisfying the *Medium* selection criterion are used as probes since for calorimeter-tagged muons with $p_T > 100$ GeV, the contribution of non-prompt muons in the OC region is not modelled correctly by the simultaneous fit.

9. Lange DJ (2001) The EvtGen particle decay simulation package. Nucl Instrum Meth 462(1–2):152–155. https://doi.org/10.1016/S0168-9002(01)00089-4

10. ATLAS Collaboration (2016) Muon reconstruction performance of the ATLAS detector in proton-proton collision data at ps = 13 TeV. Eur Phys J C 76:292

11. ATLAS Collaboration (2018). https://twiki.cern.ch/twiki/bin/view/AtlasPublic/MuonPerformancePublicPlots

12. Aaltonen T et al (2009) Search for new particles decaying into dijets in protonantiproton collisions at ps = 1.96 TeV. Phys Rev D 79:112002. https://doi.org/10.1103/PhysRevD.79.112002.

Chapter 9
Performance of Monitored Drift Tube Muon Chambers at High Rates

In 2017, the LHC exceeded its design luminosity of $\mathscr{L} = 1 \cdot 10^{34}\,\mathrm{cm}^{-2}\mathrm{s}^{-1}$ [1] by a factor of two (cf. Fig. 5.3b). The LHC experiments need to cope with this increased instantaneous luminosity in terms of rate capability, radiation hardness and trigger selectivity in order to guarantee a successful physics program. After LHC Run 3, an even more ambitious luminosity upgrade to $\mathscr{L} = 7 \cdot 10^{34}\,\mathrm{cm}^{-2}\mathrm{s}^{-1}$ will take place ushering the era of the High-Luminosity LHC (HL-LHC) [2]. In the course of the associated detector upgrades, a significant number of Monitored Drift Tube (MDT) chambers is planned to be replaced. In order to predict the behaviour of the currently installed MDT chambers within the forthcoming high rate conditions, the background hit rates, spatial resolution and reconstruction efficiency of individual MDT chambers are measured using LHC Run 2 collision data.

An MDT chamber consists of several layers of pressurised aluminium drift tubes with a diameter of 29.97 mm filled with an Ar/CO_2 gas mixture (93%/7%) at a pressure of 3 bar [3]. In the centre of the tube, a tungsten-rhenium wire with a diameter of 50 μm is kept at a potential of +3080 V with respect to the tube walls. Muons traversing the chamber ionise the gas molecules and the emerging electrons (ions) drift towards the wire (tube wall) in the radial electric field. Within about 150 μm of the wire, the strong electric field allows the electrons to obtain enough kinetic energy between two collisions with other molecules in order to ionise further Argon atoms resulting in an avalanche multiplying the primary ionisation charge by a factor of $2 \cdot 10^4$ which is also referred to as gas amplification [4]. The time difference between the avalanche arrival time and the actual muon traversal of the tube can be translated into a drift radius by the so-called space-to-drift-time relationship $r(t)$ [5]. The measured drift radii in all layers of a chamber allow for the reconstruction of muon segments as described in Sect. 7.1.

© Springer Nature Switzerland AG 2019
N. M. Köhler, *Searches for the Supersymmetric Partner of the Top
Quark, Dark Matter and Dark Energy at the ATLAS Experiment*,
Springer Theses, https://doi.org/10.1007/978-3-030-25988-4_9

9.1 Estimation of Background Hit Rates

The majority of hits in the MDT chambers arises from radiation originating from secondary interactions of collision products in the beam pipe, detector components, shielding or support structure. Figure 9.1 shows the simulated photon flux in one quadrant of the ATLAS detector at the LHC design luminosity of $1 \cdot 10^{34}\,\mathrm{cm}^{-2}\mathrm{s}^{-1}$. The innermost layer of the end-cap MDT chambers (EI) is permeated by the highest photon flux and thus the highest background hit rates are expected. The same argument holds for the flux of neutrons as depicted in Fig. A.5 in the Appendix.

The expected background rates per drift tube for the HL-LHC design luminosity of $\mathscr{L} = 7 \cdot 10^{34}\,\mathrm{cm}^{-2}\mathrm{s}^{-1}$ are shown in Fig. 9.2. Hit rates of up to 655 kHz/tube are expected for the innermost EI chambers.

In order to estimate the background hit rates depending on the instantaneous luminosity from pp collision data, a dedicated dataset comprising an integrated luminosity of $4.0\,\mathrm{fb}^{-1}$ is compiled based on data taken in the years 2016 and 2017 as summarised in Table A.3 in the Appendix. The background hit rate of a given MDT chamber is estimated by retrieving all hits recorded in the chamber for all events where no muon was reconstructed and extrapolated to the chamber. Since in this case all hits are expected to be background hits, the average height of the spectrum of recorded drifttimes divided by the number of tubes per chamber and

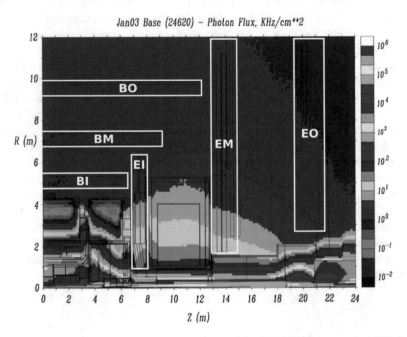

Fig. 9.1 Simulated flux of photons in one quadrant of the ATLAS detector at the LHC design luminosity of $1 \cdot 10^{34}\,\mathrm{cm}^{-2}\mathrm{s}^{-1}$. The inner (I), middle (M) and outer (O) layers of muon chambers in barrel (B) and end-caps (E) are indicated [4]

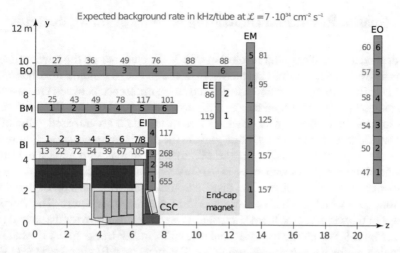

Fig. 9.2 Extrapolated background hit rates in the ATLAS muon chambers per drift tube based on measurements performed at $\sqrt{s} = 8\,\text{TeV}$ in 2012 and extrapolated to the HL-LHC design luminosity of $\mathscr{L} = 7 \cdot 10^{34}\,\text{cm}^{-2}\text{s}^{-1}$ [4]

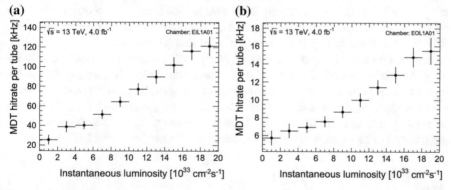

Fig. 9.3 Background hit rate per tube estimated from the average height of the spectrum of recorded drifttimes for one of the one of the innermost EI (**a**) and EO (**b**) chambers, respectively

the number of events considered yields the background hit rate. Figure 9.3 shows the estimated background hit rate in one of the innermost EI chambers as well as one of the innermost EO chambers. For the innermost EI chamber, the measured background hit rate is 120 kHz per tube while the predicted value is 187 kHz per tube (cf. Fig. 9.2).

9.2   Spatial Resolution Depending on Background Hit Rates

The dependence of the spatial resolution of a MDT chamber on the background hit rate can be estimated by studying the width of the so-called *residual* distribution. As illustrated in Fig. 9.4(a), a residual is defined as the difference between the measured drift radius $r(t)$ inside a tube and the minimum distance between the fitted muon track and the tube wire d(wire, track). In order to obtain an unbiased value of the residual, the track fit is rerun excluding hist in the tube layer in question. Figure 9.4b shows the residual distribution for the ME track of muons satisfying the *Medium* identification criterion as well as $p_T > 20$ GeV and $|\eta| < 2.5$ (cf. Table A.4) for one of the innermost EI chambers. The distribution is fitted with the sum of two Gaussian functions which are both constrained to be centred around zero. The weighted average of the widths of the fitted Gaussians is taken as a measure for the spatial resolution. Figure 9.5(a) shows the measured resolution of ME tracks for all EI chambers depending on the instantaneous luminosity. The innermost EI chambers (EI1) show a worsening spatial resolution for an increasing instantaneous luminosity. The corresponding statistical uncertainties on the spatial resolution are below 2% for all but the highest instantaneous luminosity bins (cf. Fig. A.6 in the Appendix). The decrease of spatial resolution with increasing instantaneous luminosity is not observed for the middle end-cap chambers (EM) depicted in Fig. 9.5(b) since in this region of the ATLAS cavern less background radiation is present (cf. Fig. 9.1).

Figure 9.6 shows the spatial resolution of all MDT chambers located in the innermost layer of the barrel (BI) depending on the instantaneous luminosity. While the layers BI1–BI5 do not show any dependence of the spatial resolution on the instantaneous luminosity, the chambers closer to the end-cap region (BI6–BI7) suffer from a worsening spatial resolution with increasing instantaneous luminosity. The BI7 layers are planned to be replaced by small-diameter MDT chambers after LHC Run 2 [6].

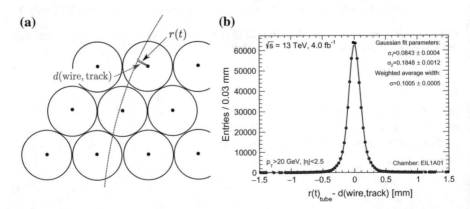

Fig. 9.4 Illustration of the calculation of MDT residuals from the drift radius $r(t)$ and the distance between the MDT wire and the fitted muon track (**a**) as well as the residual distribution with the double Gaussian fit for one of the innermost EI chambers (**b**)

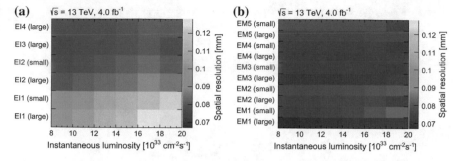

Fig. 9.5 Spatial resolution depending on the instantaneous luminosity for the different layers of EI (**a**) and EM (**b**) MDT chambers. The measurement combines the information of all MDT chambers of the same type for a given layer. EI1–EI4 (EM1–EM5) denotes the distance of the chamber layer with respect to the beam pipe. Small and large chambers are shown separately

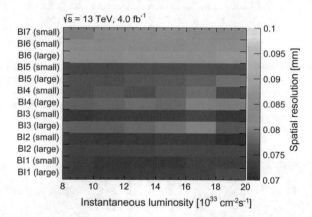

Fig. 9.6 Spatial resolution depending on the instantaneous luminosity for the different layers of the innermost layer of barrel chambers. The measurement combines the information of all MDT chambers of the same type for a given layer in $\eta > 0$. BI1–BI7 denotes the distance of the barrel chamber layer with respect to the transverse plane. Small and large chambers are shown separately

The spatial resolution also depends on the radius at which a muon traverses the drift tube since for small radii, the time difference between the first and the last electrons reaching the wire and needed to pass the readout threshold, is larger and thus, results in a worse spatial resolution [4]. For low background hit rates, the spatial resolution improves with increasing drift distance of the electrons until it saturates at $r \gtrsim 8$ mm (cf. Fig. 9.7). Increasing background hit rates worsen the spatial resolution for $r \gtrsim 6$ mm due to space charge effects caused by the large number of ions created by previous background hits [4]. Additionally, for small radii, the presence of high background hit rates and associated space charge effects reduces the gas amplification resulting in worse discrimination of a muon signal from electronic noise and a worsening spatial resolution.

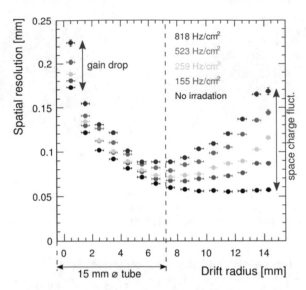

Fig. 9.7 Radial dependence of the MDT spatial resolution for different photon irradiation rates [7]. The degradation of the resolution with increasing background flux due to space charge fluctuations is limited to the region $r \gtrsim 6$ mm and increases with the drift radius. Loss of gas amplification (gain) due to shielding of the wire potential worsens the discrimination between muon signals and electronic noise resulting in a decreasing spatial resolution at small radii [4]

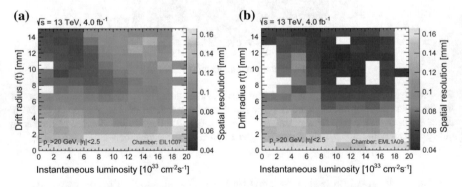

Fig. 9.8 Spatial resolution shown against the instantaneous luminosity and the measured drift radius for one of the innermost EI (**a**) and EM (**b**) chamber, respectively. The residual distribution has to contain at least 20 entries and the uncertainty on the average width of the Gaussian fit has to be less than 50% in order to be shown

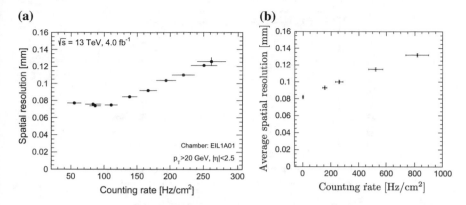

Fig. 9.9 Average spatial resolution as a function of the background counting rate measured in one of the innermost EI chambers (**a**) as well as in a test setup located at the GIF at CERN in 2003 and 2004 (**b**) [11]. For (**a**), the resolution is obtained from the width of the central Gaussian used in the double Gaussian fit

Figure 9.8(a) shows the spatial resolution of one of the innermost EI chambers drawn against the instantaneous luminosity and the measured drift radius. For increasing instantaneous luminosities, a worse spatial resolution for high drift radii is observed as expected from test beam measurements performed at the Gamma Irradiation Facility (GIF) [8] at CERN (cf. Fig. 9.7). The decreasing spatial resolution for radii $r \gtrsim 6$ mm under high background hit rates is one reason for the replacement of inner end-cap MDT chambers by different detector technologies [9, 10]. However, the MDT chambers located in the middle end-cap (EM) layer will not be replaced. Figure 9.8(b) shows the spatial resolution of one of the innermost EM chambers drawn against the instantaneous luminosity and the measured drift radius. The spatial resolution is independent of the instantaneous luminosities up to $\mathscr{L} = 1.8 \cdot 10^{34}$ cm^{-2}s^{-1}.

The measured background hit rate per tube can be translated into the counting rate per area by dividing the hit rate estimated in Sect. 9.1 by the tube diameter times the average tube length. Figure 9.9(a) depicts the spatial resolution averaging over the drift radii in dependence of the counting rate per area for one of the innermost chambers of EI. The result is comparable with measurements performed in 2003 and 2004 at the GIF at CERN (cf. Fig. 9.9b).

9.3 Chamber Efficiency Depending on Background Hit Rates

In order to study the dependence of the muon reconstruction efficiency on the background hit rate, a chamber reconstruction efficiency is estimated for each MDT chamber. Calorimeter-tagged muons with $p_T > 20$ GeV and $|\eta| < 2.5$ expected to

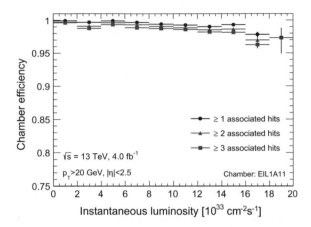

Fig. 9.10 Chamber efficiency depending on the instantaneous luminosity for one of the innermost end-cap chambers. The efficiencies for requiring one, two or three MDT hits inside the chamber which are associated to the fitted track of the *Medium* muon are shown. In addition to the selection described in Table A.4, additional requirements are made in order to suppress efficiency losses due to regions close to the chamber borders and the two-layer requirement of the *Medium* identification algorithm (cf. Appendix A.2.1)

traverse the chamber of interest are matched to reconstructed muons passing the *Medium* identification criterion which need to have at least one hit in this chamber associated to the muon track. A *Medium* muon is considered to be matched to the calorimeter tagged muon in case its track is located within $\Delta R < 0.05$ from the calorimeter-tagged muon. Requiring that the *Medium* muon did not cause the event to be recorded ensures that the measurement is unbiased with respect to the trigger system. The requirements applied for the estimation of the chamber efficiencies are summarised in Table A.4. Figure 9.10 shows the estimated chamber efficiency for one of the innermost chambers of the inner end-cap layer depending on the instantaneous luminosity. Besides the requirement that the *Medium* muon has to have at least one associated MDT hit inside the chamber, also the resulting efficiencies for requiring at least two or three hits, respectively, are shown. The chamber reconstruction efficiencies are independent of the instantaneous luminosity up to roughly $\mathscr{L} = 1 \cdot 10^{34}\,\mathrm{cm}^{-2}\mathrm{s}^{-1}$, the LHC design-luminosity, which corresponds to a background hit rate of approximately $80\,\mathrm{kHz}$/tube (cf. Fig. 9.3a). For higher background rates, the chamber efficiency decreases by 5% decrease at $\mathscr{L} = 2 \cdot 10^{34}\,\mathrm{cm}^{-2}\mathrm{s}^{-1}$.

Figure 9.11(a) shows the relative decrease of the chamber efficiency depending on the instantaneous luminosity with respect to the lowest instantaneous luminosity bin for all EI chambers grouped by their size and distance to the beam pipe. All innermost EI chambers (EI1) show a similar trend as described for Fig. 9.10. Looking at the chamber efficiency for the middle end-cap chambers, no efficiency drop is observed up to instantaneous luminosities of $\mathscr{L} = 2 \cdot 10^{34}\,\mathrm{cm}^{-2}\mathrm{s}^{-1}$ (cf. Fig. 9.11b).

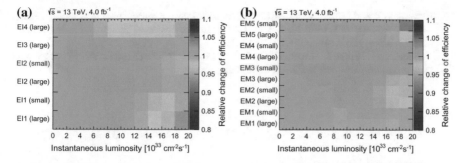

Fig. 9.11 Relative change of chamber efficiencies depending on the instantaneous luminosity for the different layers of EI (**a**) and EM (**b**) MDT chambers. The measurement combines the information of all MDT chambers of the same type for a given layer. EI1–EI4 (EM1–EM5) denotes the distance of the chamber layer with respect to the beam pipe. Small and large chambers are shown separately

The corresponding statistical uncertainties on the relative change of the combined chamber efficiencies are below 1% for all but the highest instantaneous luminosity bins (cf. Fig. A.9 in the Appendix).

References

1. Baranov S et al (2005) Estimation of radiation background, impact on detectors, activation and shielding optimization in ATLAS
2. Schmidt B (2016) The high-luminosity upgrade of the LHC: physics and technology challenges for the accelerator and the experiments. J Phys Conf Ser 706(2):022002. https://doi.org/10.1088/1742-6596/706/2/022002
3. ATLAS Collaboration (2008) The ATLAS experiment at the CERN large hadron collider. JINST 3:S08003. https://doi.org/10.1088/1748-0221/3/08/S08003
4. Schwegler P, Kroha H (2014) High-rate performance of muon drift tube detectors. Presented 14 June 2014. https://cds.cern.ch/record/1746370
5. Horvat S, Kortner O, Kroha H (2006) Determination of the spatial drift-tube resolution using muon tracks
6. Aielli G et al (2016) The ATLAS BIS78 Project. Technical report ATL-MUON-INT- 2016-002. Geneva: CERN, June 2016. https://cds.cern.ch/record/2161109
7. Horvat S et al (2006) Operation of the ATLAS muon drift-tube chambers at high background rates and in magnetic fields. IEEE Trans Nucl Sci 53:562–566. https://doi.org/10.1109/TNS.2006.872636
8. Agosteo S et al (2000) A facility for the test of large-area muon chambers at high rates. Nucl Instrum Methods Phys Res Sect A: Accel, Spectrometers, Detect Assoc Equip 452(1):94–104. ISSN: 0168-9002
9. Stelzer B (2016) The new small wheel upgrade project of the ATLAS Experiment. Nucl Part Phys Proc 273–275. In: 37th International conference on high energy physics (ICHEP), pp 1160–1165. https://doi.org/10.1016/j.nuclphysbps.2015.09.182. ISSN: 2405-6014

10. Kawamoto T et al (2013) New small wheel technical design report. Technical report CERN-LHCC-2013-006. ATLAS-TDR-020. ATLAS New Small Wheel Technical Design Report. June 2013. https://cds.cern.ch/record/1552862
11. Dubbert J et al (2016) Performance of the ATLAS muon drift-tube chambers at high background rates and in magnetic fields. arXiv:1604.01598 [physics.ins-det]

Part IV
Searches for New Particles Decaying into Jets and Missing Transverse Energy

Chapter 10
Searches for New Particles

Searches for new particles at the LHC are relying on the intrinsically random nature of quantum mechanics manifested in the indeterminism about the outcome of the particle collisions. The production cross section of a particle is predicting its expected production rate in the LHC pp collisions (cf. Eq. (5.2)). Thus, from one single collision one cannot infer any significant result, but a statistical interpretation of all the recorded collision data needs to be performed which is heavily relying on the comparison to simulated data of all physics processes involved in the search. Consequently, for achieving the best possible sensitivity in the search for new particles, the following aspects have to be carefully considered:

- The underlying theoretical model postulating a new particle is also predicting its production *cross section*. Since the production cross section is determining the expected production rate, it should be of a reasonable size compared to the cross sections of the SM particles which are produced in the majority pp collisions. Thus, a production cross section as large as possible is favoured, however, since we assume cross sections to be predetermined by the fundamental laws of the universe, we cannot tune the cross section of a certain potential particle, but at least chose the theoretical model we are looking for in the best possible way.
- Since the expected number of events in which new particles may be produced at the LHC scales with the *integrated luminosity* (cf. Eq. (5.3)), the sensitivity of the search for a new particle increases with the integrated luminosity and thus, the size of the dataset recorded by the experiment. Large integrated luminosities can be achieved by finding the optimal trade-off between increasing the duration of data taking and maximising the instantaneous luminosity. The latter is limited by the capabilities of the detector elements to operate at very high rates while the former is a question of how long the LHC and its experiments are supposed to be maintained before providing valuable physics results.
- Due to resolution or efficiency effects in both detector elements and reconstruction algorithms, some degree of randomness is introduced to the recorded collision data.

© Springer Nature Switzerland AG 2019

N. M. Köhler, *Searches for the Supersymmetric Partner of the Top Quark, Dark Matter and Dark Energy at the ATLAS Experiment*, Springer Theses, https://doi.org/10.1007/978-3-030-25988-4_10

In order to precisely predict the trajectories and energy deposits of the particles created in the collision which is also referred to as the event topology, *Monte Carlo simulation* is exploited which allows to take into account the full complexity of the production and detection of all physical processes.

This part of the thesis introduces three searches for physics beyond the SM in final states with jets and missing transverse energy based on $36.1 \, \text{fb}^{-1}$ of pp collision data taken in 2015 and 2016 at $\sqrt{s} = 13 \, \text{TeV}$. A detailed description of the theoretical model used for the simulation of the beyond SM process of interest will be given at the beginning of each Chapter describing a search. Since the simulation of the SM processes as well as the general strategy of searches for new particles performed by the ATLAS collaboration is common to all searches, they are introduced in this chapter. Chapter 11 presents a search for the supersymmetric partner of the top quark as well as the lightest neutralino which can be a Dark Matter candidate (cf. Sect. 4.4). Subsequently, a search for WIMP-like Dark Matter is presented in Chap. 12 which does not presume the existence of Supersymmetry, although its phenomenology shares several similarities with the search for top squarks. Chapter 13 finally presents a search for Dark Energy assuming an effective field theory predicting light scalar Dark Energy particles which can be potentially produced at the LHC (cf. Sect. 3.4).

10.1   Simulation of Standard Model Processes

For every search for physics beyond the SM, the SM processes have to be precisely understood in order to estimate their contributions in the regions of phase space where the new physics processes are expected to emerge. This Chapter introduces the simulation of SM processes relevant for searches at the LHC using Monte Carlo methods.

Figure 10.1 shows a summary of the production cross section measurements of SM particles in pp collisions performed by the ATLAS collaboration with the LHC Run 1 and Run 2 datasets. The most dominating SM processes at the LHC, the so-called multi-jet production arising from various processes of the strong interaction described by the QCD, is not included in the Figure. The simulation of multi-jet production with sufficient statistics compared to the dataset recorded goes beyond the scope of the computational feasibility of the WLCG and thus, is usually estimated using data-driven methods. The remaining dominating processes are W and Z boson production as well as $t\bar{t}$ and single top quark production. The production of WW, WZ and ZZ is commonly referred to as diboson production. As mentioned earlier, the estimation of the SM contributions to the selected events requires Monte Carlo (MC) simulated data.

The simulation of W and Z boson production with additional jets, including jets from the hadronisation of b- and c-quarks, is commonly called W+jets (cf. Fig. 10.2) and Z+jets (cf. Fig. 10.3), respectively. W+jets and Z+jets events are generated with SHERPA 2.2.1 [2] where matrix elements are calculated for up to two

Standard Model Total Production Cross Section Measurements   Status: March 2018

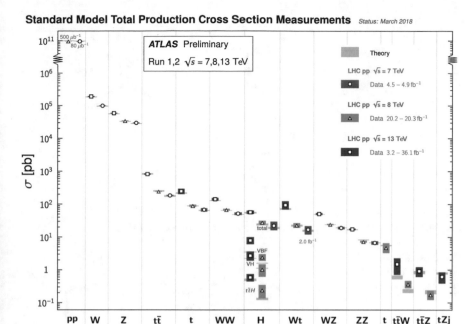

Fig. 10.1 Summary of several Standard Model total production cross section measurements, corrected for leptonic branching fractions, compared to the corresponding theoretical expectations. All theoretical expectations were calculated at NLO or higher. The luminosity used for each measurement is indicated close to the data point. Some measurements have been extrapolated using branching ratios as predicted by the Standard Model for the Higgs boson. Uncertainties for the theoretical predictions are quoted from the original ATLAS papers. They were not always evaluated using the same prescriptions for PDFs and scales. Not all measurements are statistically significant yet. The first bin shows the total pp cross section [1]

additional partons at next-to-leading order (NLO) and four partons at leading order (LO) using the COMIX [3] and OPEN LOOPS [4] matrix element generators. The NNPDF3.0NNLO [5, 6] set is used for the parton distribution functions (PDFs) and the parton showering is done with the built-in SHERPA PS [7] using the ME+PSNLO prescription [8] and a dedicated underlying-event (UE) tune developed by the SHERPA authors. Events are generated in non-overlapping slices of the vector boson p_T and the presence or absence of jets from the hadronisation of b- and c-quarks in order to maximise the MC statistics.

For the simulation of top quark pair production $t\bar{t}$ (cf. Fig. 10.4) only events where at least one top quark decays via a leptonic W boson decay are generated, since fully hadronic top quark decays ideally do not lead to missing transverse energy and do not contribute to final states with jets and missing transverse energy. SM $t\bar{t}$ production only contributes in the final state with jets and missing transverse momentum in case one W boson decays into a lepton-neutrino pair and the lepton is not reconstructed properly. SM $t\bar{t}$ production is simulated using the POWHEG- BOX 2 generator [9].

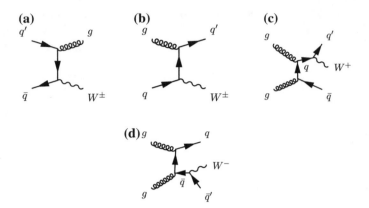

Fig. 10.2 Selection of Feynman graphs for W+jets production at the LHC

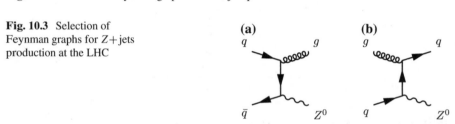

Fig. 10.3 Selection of Feynman graphs for Z+jets production at the LHC

The production of single top quarks can be categorised into electroweak t-channel production (cf. Fig. 10.5a) where two quarks (one of which is a bottom quark) are exchanging a W boson altering the flavour of the quarks resulting in the creation of a top quark. The t-channel production of top quarks has the highest production cross section but usually, the second quark is of first or second generation. Figure 10.5(b) illustrates the production of single top quarks in the s-channel where a quark and an anti-quark differing in the third component of the weak isospin are producing a virtual W boson which subsequently produces a top quark and a bottom anti-quark. The third possibility to produce a single top quark is the interaction of a quark and a gluon which produce a virtual bottom quark creating a top quark and a W boson (cf. Fig. 10.5c) which is thus referred to as Wt-channel production. t-channel and s-channel are referring to the Mandelstam variables [10] which are describing the arrangement of the four-momenta of incoming and outgoing particles in the scattering process. For each channel, events are generated using the POWHEG- BOX 2 generator. For all processes involving top quarks, a top quark mass of 172.5 GeV is assumed. The parton showering is done using PYTHIA 6.428 [11] with the PERUGIA 2012 [12] set of tuned shower and underlying-event parameters and the CT10 PDF set [13] for the matrix element calculations. The simulations are normalised to their next-to-next-to-leading order (NNLO) cross-section [14, 15] including the resummation of soft gluon emission at next-to-next-to-leading-log (NNLL) accuracy [16–18] using TOP++2.0 [19–22].

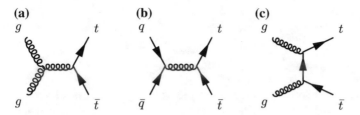

Fig. 10.4 Selection of Feynman graphs for $t\bar{t}$ production at the LHC

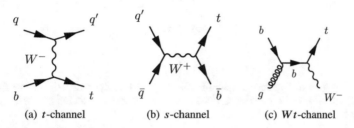

(a) *t*-channel   (b) *s*-channel   (c) *Wt*-channel

Fig. 10.5 Selection of Feynman graphs for single top quark production at the LHC

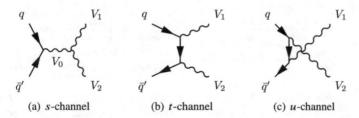

(a) *s*-channel   (b) *t*-channel   (c) *u*-channel

Fig. 10.6 Selection of Feynman graphs for diboson VV production ($V \in \{W, Z\}$) at the LHC ordered by the Mandelstam variables of the scattering process. Subfigure (a) is showing the triple gauge coupling vertex which only occurs for *s*-channel production

Diboson processes (cf. Fig. 10.6) are simulated with the SHERPA generator and the NNPDF3.0NNLO PDF set in combination with the built-in parton showering prescription and underlying-event tune. The matrix elements are calculated for up to one additional parton at NLO and up to three additional partons at LO.

Top quark pair production in association with a W or Z boson and possibly additional jets is commonly referred to as $t\bar{t} + V$ ($V \in \{W, Z\}$) (cf. Fig. 10.7). $t\bar{t} + V$ production is modelled using the aMC@NLO 2.2.3 [23] generator at NLO and PYTHIA 8.186 [24] for parton showering. The PDF set used is NNPDF3.0NNLO, the underlying-event tune A14 [25].

As becoming apparent in Sect. 11.4, the production of $t\bar{t}$ in association with a highly energetic photon γ^* can be used in order to study the normalisation of $t\bar{t} + Z$ production. Replacing the Z boson by a photon γ^* in the $t\bar{t} + Z$ diagrams (cf. Fig. 10.7), a similar event topology can be observed for high photon p_T. $t\bar{t} + \gamma^*$ events are generated using MADGRAPH 2.2.3 at NLO and showered in

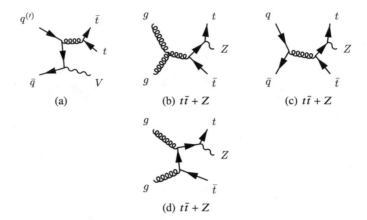

Fig. 10.7 Selection of Feynman graphs for $t\bar{t} + V$ ($V \in \{W, Z\}$) production at the LHC. Note that (b)–(d) are $t\bar{t} + Z$ diagrams only, since a Z boson is radiated on top of SM $t\bar{t}$ production (cf. Fig. 10.4)

Table 10.1 Overview of the simulations used for the SM processes. More details of the generator configurations can be found in [26–29]

Process	Generator	Showering	PDF set	UE tune	Order
$W+$jets	SHERPA 2.2.1	SHERPA 2.2.1	NNPDF3.0NNLO	SHERPA default	NLO
$Z+$jets	SHERPA 2.2.1	SHERPA 2.2.1	NNPDF3.0NNLO	SHERPA default	NLO
$t\bar{t}$	POWHEG- BOX 2	PYTHIA 6.428	CT10 (NLO)	PERUGIA 2012	NNLO+NNLL
Single top:					
• t-channel	POWHEG- BOX 1	PYTHIA 6.428	CT10 F4 (NLO)	PERUGIA 2012	NNLO+NNLL
• s-/Wt-channel	POWHEG- BOX 2	PYTHIA 6.428	CT10 (NLO)	PERUGIA 2012	NNLO+NNLL
Diboson	SHERPA 2.2.1	SHERPA 2.2.1	NNPDF3.0NNLO	SHERPA default	NLO
$t\bar{t} + W$	aMC@NLO 2.2.3	PYTHIA 8.186	NNPDF3.0NNLO	A14	NLO
$t\bar{t} + Z$	aMC@NLO 2.2.3	PYTHIA 8.186	NNPDF3.0NNLO	A14	NLO
$t\bar{t} + \gamma^*$	MADGRAPH 2.2.3	PYTHIA 8.186	NNPDF3.0NNLO	A14	NLO

PYTHIA 8.186. As for $t\bar{t} + V$ processes, the PDF set used is NNPDF3.0NNLO, the underlying-event tune A14.

The matrix element generators, parton showering prescriptions, parton distribution function sets, underlying-event tune parameters and cross-section orders for all simulated processes are summarised in Table 10.1.

All simulated events are processed through a GEANT 4 simulation [30] of the ATLAS detector [31]. This makes it possible to apply the same reconstruction algorithms to the simulated as to the recorded dataset. In order to account for the effects of additional pp collisions in the same or nearby bunch crossings, a varying number of additional simulated minimum-bias interactions is generated and overlaid onto each simulated hard-scatter event. The minimum-bias events are generated using the soft QCD processes of PYTHIA 8.186 with the A2 tune [32] and the MSTW2008LO

PDF set [33]. Afterwards, the simulated events are reweighted to match the distribution of the number of pp interactions per bunch crossing in the experimental data (cf. Sect. 6.7).

Furthermore, correction factors are applied to all simulated events in order to correct for differences between experimental and simulated data in the reconstruction and isolation efficiencies, b-tagging efficiencies and mis-tag rates, momentum scale, energy resolution and trigger efficiencies in case of lepton requirements (cf. Sect. 6.8).

10.2 General Search Strategy

Before any statistical interpretation of measured data can be performed, every search for physics beyond the SM starts as a simple classification problem: All pp collisions recorded, which are also referred to as *events*, have to be classified as either containing SM processes only, or showing features which are potentially caused by new physical processes. The former is usually referred to as SM *background* while the latter is called *signal*. The main goal of a search is to separate signal-like events from background-like events with the highest possible purity by carefully considering the impacts of the uncertainties of all parameters which enter the classification procedure.

Commonly, the classification between signal-like and background-like events is done by looking for properties of the event which differ between background and signal processes. Such properties are called *discriminating variables*. Applying cuts to a collection of several discriminating variables is called an *event selection*. A selection which enriches the data sample with signal-like events while suppressing as much background as possible is called a *signal region* (SR). Consequently, the MC simulation of signal and background processes is the essential prerequisite for finding discriminating variables and optimising the values at which a cut on the variables has to be applied in order to define the best possible SR depending on the signal process.

However, every simulation is based on certain theoretical assumptions and may model the real physics process with limited accuracy. In order to estimate the number of events for a given background process in the SR, one favours to not only rely on the simulation only, but tries to validate the agreement of the simulation with real pp collision data whenever possible. This can be done, by manipulating the selection criteria of at least one discriminating variable in order define a disjoint region of phase space which contains a negligible signal contribution and enriches the background process of interest. In case cuts are manipulated in a way that a certain background process is enriched compared to all remaining background processes, a so-called *control region* (CR) for this process is defined. Comparing the numbers of recorded data $N_{\text{Data}}^{\text{CR}}$ and simulated events $N_{\text{MC}}^{\text{CR}}$ in the CR, makes it possible to determine a transfer factor μ_b^{CR} whose application to the background b of interest yields to

$$N_{\text{Data}}^{\text{CR}} = \mu_b^{\text{CR}} \cdot N_{\text{b}}^{\text{CR}} + \left(N_{\text{MC}}^{\text{CR}} - N_{\text{b}}^{\text{CR}} \right). \tag{10.1}$$

Fig. 10.8 Schematic drawing of the ATLAS strategy for data-driven background estimation. Control regions ideally are disjoint but still close in phase space to signal regions, although there might be room in between for the definition of a validation region. In general, different control and validation regions should also be disjoint among each other

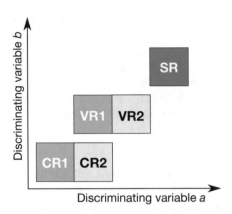

The definition of a CR allows for the estimation of the normalisation of a background process from data while taking the shape of the discriminating variables from simulation. The justification for the latter is that CRs should ideally be as close as possible in phase space to the according SR, but being disjoint to it. Furthermore, it is common to verify that the estimated scaling factor μ_b^{CR} does not depend significantly on the inverted variables for the CR definition by defining so-called *validation regions* (VRs) which are located in between CR and SR being disjoint to both of them (cf. Fig. 10.8). In case the agreement between recorded data and simulation gets significantly worse in the VR, the CR definition has to be reconciled.

The rigorous requirements for sensible CRs sometimes do not permit to define a CR for every background process. In particular processes leading to the very same signature as a signal process inside the detector make the definition of a CR almost impossible. Thus, usually those types of background processes, which are also referred to as *irreducible background* processes are fully estimated from MC simulation and thus require the most precise MC modelling possible.

References

1. ATLAS public results - Standard model summary plots (2018). https://atlas.web.cern.ch/Atlas/GROUPS/PHYSICS/CombinedSummaryPlots/SM/
2. Gleisberg T et al (2009) Event generation with SHERPA 1.1. JHEP 2009(02):007. https://doi.org/10.1088/1126-6708/2009/02/007
3. Gleisberg T, Höche S (2008) Comix, a new matrix element generator. JHEP 2008(12):039. https://doi.org/10.1088/1126-6708/2008/12/039
4. Cascioli F, Maierhofer P, Pozzorini S (2012) Scattering amplitudes with open loops. Phys Rev Lett 108:111601. https://doi.org/10.1103/PhysRevLett.108.111601
5. Ball RD et al (2013) Parton distributions with LHC data. Nucl Phys B 867:244–289. https://doi.org/10.1016/j.nuclphysb.2012.10.003, arXiv:1207.1303 [hep-ph]
6. Catani S et al (2009) Vector boson production at hadron colliders: a fully exclusive QCD calculation at NNLO. Phys Rev Lett 103:082001. https://doi.org/10.1103/PhysRevLett.103.082001

7. Schumann S, Krauss F (2008) A parton shower algorithm based on Catani-Seymour dipole factorisation. JHEP 2008(03):038. https://doi.org/10.1088/1126-6708/2008/03/038
8. Höche S et al (2013) QCD matrix elements + parton showers: The NLO case. JHEP 2013(04):027. https://doi.org/10.1007/JHEP04(2013)027
9. Alioli S et al (2010) A general framework for implementing NLO calculations in shower Monte Carlo programs: the POWHEG BOX. JHEP 2010(06):043. https://doi.org/10.1007/JHEP06(2010)043, arXiv:1002.2581 [hep-ph]
10. Mandelstam S (1958) Determination of the pion-nucleon scattering amplitude from dispersion relations and unitarity. General theory. Phys Rev 112(4):1344–1360. https://doi.org/10.1103/PhysRev.112.1344
11. Sjöstrand T, Mrenna S, Skands P (2006) PYTHIA 6.4 physics and manual. JHEP 2006(05):026. https://doi.org/10.1088/1126-6708/2006/05/026
12. Skands PZ (2010) Tuning Monte Carlo generators: the Perugia tunes. Phys Rev D 82:074018. https://doi.org/10.1103/PhysRevD.82.074018
13. Lai H-L et al (2010) New parton distributions for collider physics. Phys Rev D 82:074024. https://doi.org/10.1103/PhysRevD.82.074024
14. Czakon M, Mitov A (2013) NNLO corrections to top pair production at hadron colliders: the quark-gluon reaction. JHEP 2013(01):080. https://doi.org/10.1007/JHEP01(2013)080
15. Czakon M, Mitov A (2012) NNLO corrections to top-pair production at hadron colliders: the all-fermionic scattering channels. JHEP 2012(12):054. https://doi.org/10.1007/JHEP12(2012)054
16. Czakon Mi, Fiedler P, Mitov A (2013) Total top-quark pair-production cross section at hadron colliders through $O(\alpha_S^4)$. Phys Rev Lett 110:252004. https://doi.org/10.1103/PhysRevLett.110.252004
17. Bärnreuther P, Czakon M, Mitov A (2012) Percent level precision physics at the tevatron: first genuine NNLO QCD corrections to $q\bar{q} \to t\bar{t} + X$. Phys Rev Lett 109:132001. https://doi.org/10.1103/PhysRevLett.109.132001
18. Cacciari M et al (2012) Top-pair production at hadron colliders with next-tonext-to-leading logarithmic soft-gluon resummation. Phys Lett B 710:612. https://doi.org/10.1016/j.physletb.2012.03.013
19. Kidonakis N (2011) Next-to-next-to-leading-order collinear and soft gluon corrections for t-channel single top quark production. Phys Rev D 83:091503. https://doi.org/10.1103/PhysRevD.83.091503, arXiv:1103.2792 [hep-ph]
20. Kidonakis N (2010) NNLL resummation for s-channel single top quark production. Phys Rev D 81:054028. https://doi.org/10.1103/PhysRevD.81.054028
21. Kidonakis N (2010) Two-loop soft anomalous dimensions for single top quark associated production with a W- or H-. Phys Rev D 82:054018. https://doi.org/10.1103/PhysRevD.82.054018
22. Czakon M, Mitov A (2014) Top++: a program for the calculation of the top-pair cross-section at hadron colliders. Comput Phys Commun 185:2930. https://doi.org/10.1016/j.cpc.2014.06.021
23. Alwall J et al (2014) The automated computation of tree-level and next-to-leading order differential cross sections, and their matching to parton shower simulations. JHEP 2014(07):079. https://doi.org/10.1007/JHEP07(2014)079
24. Sjöstrand T, Mrenna S, Skands PZ (2008) A brief introduction to PYTHIA 8.1. Comput Phys Commun 178:852–867. https://doi.org/10.1016/j.cpc.2008.01.036
25. ATLAS Collaboration (2014) ATLAS Pythia 8 tunes to 7 TeV data. ATL-PHYSPUB-2014-021. https://cds.cern.ch/record/1966419
26. ATLAS Collaboration (2016) Simulation of top-quark production for the ATLAS experiment at ps = 13 TeV. ATL-PHYS-PUB-2016-004. https://cds.cern.ch/record/2120417
27. ATLAS Collaboration (2016) Monte Carlo generators for the production of aW or Z/γ^* boson in association with jets at ATLAS in Run 2. ATL-PHYS-PUB- 2016-003. https://cds.cern.ch/record/2120133
28. ATLAS Collaboration (2016) Multi-boson simulation for 13 TeV ATLAS analyses. ATL-PHYS-PUB-2016-002. https://cds.cern.ch/record/2119986

29. ATLAS Collaboration (2016) Modelling of the ttH and ttV(V = W, Z) processes for ps = 13 TeV ATLAS analyses. ATL-PHYS-PUB-2016-005. https://cds.cern.ch/record/2120826
30. Agostinelli S et al (2003) GEANT4: a simulation toolkit. Nucl Instrum Meth A 506:250. https://doi.org/10.1016/S0168-9002(03)01368-8
31. ATLAS Collaboration (2010) The ATLAS simulation infrastructure. Eur Phys J C 70:823. https://doi.org/10.1140/epjc/s10052-010-1429-9
32. ATLAS Collaboration (2012) Summary of ATLAS Pythia 8 tunes. ATL-PHYS-PUB- 2012-003. https://cds.cern.ch/record/1474107
33. Martin AD et al (2009) Parton distributions for the LHC. Eur Phys J C 63:189–285. https://doi.org/10.1140/epjc/s10052-009-1072-5

Chapter 11
The Search for the Light Top Squark

As discussed in Sect. 4.3, the lighter mass eigenstate of the superpartner of the top quark, the top squark $\tilde{t}_1$, serves as a solution for the Hierarchy problem, if its mass is about 1 TeV. In all supersymmetric models considered in the scope of this thesis, R-parity is assumed to be conserved enforcing the $\tilde{t}_1$ to be produced in pairs and decay into the stable neutralino $\tilde{\chi}_1^0$ and additional SM particles. Figure 11.1 shows three different decay scenarios of the top squark pair. In the case of $m_{\tilde{t}_1} > m_t + m_{\tilde{\chi}_1^0}$, the top squark can decay into an on-shell top quark t and the lightest neutralino $\tilde{\chi}_1^0$ (cf. Fig. 11.1a), while in the case of $m_{\tilde{t}_1} > m_{\tilde{\chi}_1^\pm} > m_W + m_{\tilde{\chi}_1^0}$, the top squark can decay into the lighter chargino $\tilde{\chi}_1^\pm$ which then subsequently decays into a W boson and the lightest neutralino $\tilde{\chi}_1^0$ (cf. Fig. 11.1b). If the top squark mass is chosen to be below the sum of the top quark and the neutralino mass, the three-body decay $\tilde{t}_1 \rightarrow W + b + \tilde{\chi}_1^0$ becomes kinematically allowed (cf. Fig. 11.1c).

During LHC Run 1, searches for direct top squark pair production were performed using 20 fb^{-1} of pp collision data at $\sqrt{s} = 8$ TeV by both the ATLAS [1–5] and CMS [6–8] collaborations. No sign for the existence of top squarks was found. A summary of the exclusion limits at 95% CL derived from various ATLAS searches targeting different experimental signatures is shown in Fig. 11.2. The exclusion limits above the dashed line $\Delta m(\tilde{t}_1, \tilde{\chi}_1^0) < m_b + m_W$ correspond to the decay scenarios shown in Fig. 11.1. The dashed line $\Delta m(\tilde{t}_1, t) < m_{\tilde{\chi}_1^0}$ separates the three-body decay scenario (cf. Fig. 11.1c) from the others. The small triangle at $m_{\tilde{t}_1} \sim m_t$ was excluded by a dedicated search [4] which will not be discussed within the scope of this thesis. Top squark masses up to 700 GeV are excluded for neutralino masses up to 100 GeV. Neutralino masses up to 200 GeV are excluded for top squark masses between 400 and 680 GeV. However, the existence of a top squark with $m_{\tilde{t}_1} - m_{\tilde{\chi}_1^0} \sim m_t$ as well as $m_{\tilde{t}_1} > 700$ GeV is still possible. Furthermore, for three-body decays, the exclusion limits derived from LHC Run 1 only range up to top squark masses of 300 GeV and neutralino masses of roughly 150 GeV.

© Springer Nature Switzerland AG 2019
N. M. Köhler, *Searches for the Supersymmetric Partner of the Top Quark, Dark Matter and Dark Energy at the ATLAS Experiment*, Springer Theses, https://doi.org/10.1007/978-3-030-25988-4_11

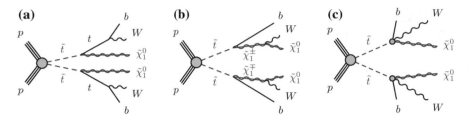

Fig. 11.1 Sketched diagrams for direct $\tilde{t}_1$ pair production and possible subsequent decays for $m_{\tilde{t}_1} > m_W + m_b + m_{\tilde{\chi}_1^0}$. Note that all graphs result in the same final state

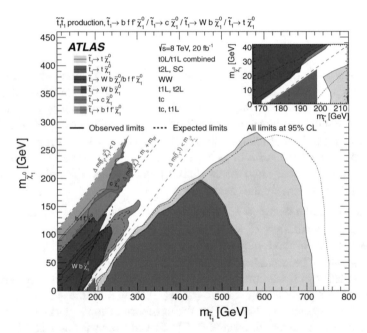

Fig. 11.2 Summary of dedicated ATLAS searches for top squark pair production based on 20 fb^{-1} of pp collision data taken at $\sqrt{s} = 8$ TeV. Exclusion limits at 95% CL are shown in the $\tilde{t}_1$-$\tilde{\chi}_1^0$ mass plane. The dashed and solid lines show the expected and observed limits, respectively, including all uncertainties except the theoretical signal cross section uncertainty (PDF and scale). Four decay modes are considered separately with 100% $\mathcal{BR}$: $\tilde{t}_1 \rightarrow t + \tilde{\chi}_1^0$ (where the $\tilde{t}_1$ is mostly right-handed), $\tilde{t}_1 \rightarrow W + b + \tilde{\chi}_1^0$ (3-body decay for $m_{\tilde{t}_1} < m_t + m_{\tilde{\chi}_1^0}$), $\tilde{t}_1 \rightarrow c + \tilde{\chi}_1^0$ and $\tilde{t}_1 \rightarrow f + f' + \tilde{\chi}_1^0$ (4-body decay). The latter two decay modes are superimposed. The region $m_{\tilde{t}_1} < 100$ GeV has not been considered for the 4-body decay [5]

11.1   Simulation of Top Squark Pair Production

For the modelling of top squark pair production, so-called *simplified models* [9, 10] are used where only one or two step (cf. Fig. 11.1b) decays are allowed. Other supersymmetric particles than $\tilde{t}_1$, $\tilde{\chi}_1^0$ and (depending on the decay scenario) $\tilde{\chi}_1^\pm$ are assumed to be decoupled to higher energies not accessible at the LHC and thus, they do not take part in any interaction or decay. To cover the full non-excluded $\tilde{t}_1$-$\tilde{\chi}_1^0$ mass plane, a grid of signal simulations is generated using a grid spacing of 50 GeV and assuming a maximal mixing between the superpartners of the left- and right-handed top quark. The signal events are simulated using the MADGRAPH5_aMC@NLO 2.2-2.4 [11] generator with PYTHIA 8 for the parton showering. The hadronisation of b- and c-hadron decays is performed with EVTGEN 1.2.0 [12]. The matrix element calculation is carried out at tree level with the emission of up to two additional partons. The matrix element is matched to parton showers using the CKKW-L [13] prescription with a matching scale set to one quarter of $m_{\tilde{t}_1}$. The PDF set used is NNPDF2.3NNLO [14], the underlying-event tune A14. All signal cross sections are calculated to NLO in the strong coupling constant, the resummation of soft-gluon emission is added at NLL accuracy [15–17]. The cross section and its uncertainty are calculated from an envelope of cross section predictions using different PDF sets and factorisation and renormalisation scales [18].

The simulated signals are processed through the ATLAS detector simulation based on GEANT 4 or through a fast simulation framework where the showers in the electromagnetic and hadronic calorimeters are simulated with a parametrised description [19]. The fast simulation framework was validated against the full GEANT 4 simulation for several selected signal samples (cf. Appendix B.2) and subsequently used for all signal processes.

11.2 Basic Experimental Signature

Depending on the decay of the W boson, the experimental signature of all signal processes shown in Fig. 11.1 can contain two, one or zero leptons. The case of zero leptons (cf. Fig. 11.3) has the highest branching fraction and is referred to as fully hadronic decay. The experimental signature are jets and missing transverse energy. Top squark pairs decaying into one or two leptons are not discussed within the scope of this thesis.

To reject events with leptons in the final states with high efficiency, the lepton identification algorithms with the highest efficiencies are chosen since keeping events with an unidentified lepton is resulting in less reliable results than rejecting events containing potential fake leptons.[1] Thus, electron candidates are selected using the *VeryLoose* identification criteria, a minimum transverse energy of $E_T > 7$ GeV and a pseudorapidity of $|\eta| < 2.47$. Also electrons in the transition region ($1.37 < |\eta| < 1.52$) between the barrel and the endcap sector of the electromagnetic calorimeter, which are satisfying the *VeryLoose* identification criteria, are considered as electron candidates. To ensure the spatial separation of electrons

---

[1] In case the electron or muon reconstruction algorithms are reconstructing a lepton although there was no lepton produced in the hard process or the subsequent decay chain, the lepton is commonly referred to as a *fake* lepton.

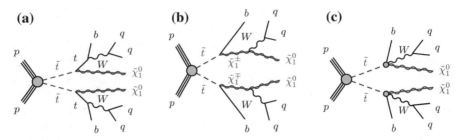

Fig. 11.3 Selection of schematic Feynman graphs for direct $\tilde{t}_1$ production and possible subsequent fully hadronic decays for $m_{\tilde{t}_1} > m_W + m_b + m_{\tilde{\chi}_1^0}$. Note that all graphs result in the same final state

with respect to other physics objects, cuts on two discriminating isolation variables are imposed. A variable-cone track-based isolation variable, $p_T^{\text{varcone20}}$, is defined as the sum of transverse momenta within a cone of $\Delta R = \min\left(\frac{10}{E_T\,[\text{GeV}]}, 0.2\right)$ around the candidate electron track. The tracks associated to the electron are excluded [20]. A calorimeter-based isolation variable, E_T^{cone20}, is defined as the sum of transverse energies of topological clusters, calibrated at the electromagnetic scale, within a cone of $\Delta R = 0.2$ around the candidate electron cluster. The transverse energy contained in a rectangular cluster of size $\Delta\eta \times \Delta\phi = 0.125 \times 0.175$ centred around the electron cluster is subtracted correcting for energy leakage of the electron outside this cluster [20]. Based on these two discriminating variables, a selection on the ratios $p_T^{\text{varcone20}}/E_T$ and E_T^{cone20}/E_T, designed to be 99% efficient for electrons from Z-boson decays, is applied as isolation criterion.

Muon candidates are selected using the *Loose* identification criteria, a minimum transverse momentum of $p_T > 6\,\text{GeV}$ and a pseudorapidity of $|\eta| < 2.7$. As for the electrons, a variable-cone track-based isolation variable, $p_T^{\text{varcone30}}$, and a calorimeter-based isolation variable, E_T^{cone20}, is defined. The track-based isolation variable, $p_T^{\text{varcone30}}$, is defined as the scalar sum of all tracks with $p_T > 1\,\text{GeV}$ in a cone of $\Delta R = \min\left(\frac{10}{p_T\,[\text{GeV}]}, 0.3\right)$ around the muon transverse momentum excluding the muon track. The calorimeter-based isolation variable, E_T^{cone20}, is defined as for electrons with an additional correction for pile-up effects [21]. Again, a selection on the ratios $p_T^{\text{varcone30}}/E_T$ and E_T^{cone20}/E_T, designed to be 99% efficient for muons from Z-boson decays, is applied as isolation criterion.

After the removal of reconstruction ambiguities between leptons and jets, only jets within a pseudorapidity of $|\eta| < 2.8$ are kept [22].

For the CRs containing leptons (cf. Sect. 11.4) the leptons have to satisfy more stringent requirements in order to reject non-prompt leptons with an higher efficiency. Electrons must satisfy the *Tight* identification criteria and have a transverse energy of $E_T > 25\,\text{GeV}$. By restricting the impact parameter requirements to $|z_0| \cdot \sin(\theta) < 0.5\,\text{mm}$ and $|d_0|/\sigma_{d_0} < 5$, electrons are constrained to originate from the primary vertex. Muons have to pass the *Medium* identification criteria and satisfy $p_T > 20\,\text{GeV}$. By restricting the impact parameter requirements to

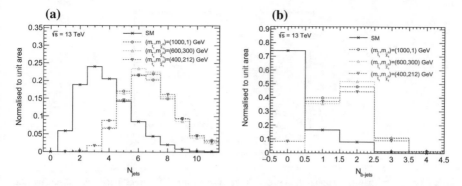

Fig. 11.4 Relative number of simulated jets and b-jets after applying a lepton veto and requiring the scalar sum of transverse momenta of all jets, H_T, to be $H_T > 150$ GeV. A selection of top squark signal processes not excluded by LHC Run 1 as well as the sum of the SM contributions apart from multijet production are shown

$|z_0| \cdot \sin(\theta) < 0.5$ mm and $|d_0|/\sigma_{d_0} < 3$, muons are constrained to originate from the primary vertex.

For the control region containing a photon, photons are selected using the *Tight* identification criteria and requiring $p_T > 150$ GeV. In order to ensure the spatial separation of photons to neighbouring objects, the calorimeter-based isolation variable with a cone of $\Delta R = 0.4$, E_T^{cone40}, is required to fulfil E_T^{cone40} [GeV] $< 0.022 \cdot p_T$ [GeV] $+ 2.45$ GeV and the track-based isolation variable with a cone of $\Delta R = 0.2$, E_T^{cone20}, is required to have a value $E_T^{cone20} < 0.065 \cdot p_T$ [23].

To define the basic experimental signature for fully hadronic decays of top squark pairs, one examines the relative distribution of the number of simulated jets and b-jets after applying a lepton veto shown in Fig. 11.4 for a selection of top squark signal processes not excluded by LHC Run 1 data. Independently of the top squark and neutralino masses, more than 95% of the simulated signal events have at least four jets while roughly half of all SM contributions apart from multijet production have less than four jets. Requiring at least one b-jet will result in the loss of up to 10% of the simulated signal events. Yet, the requirement suppresses a significant amount of SM processes not containing jets originating from b-hadrons. Thus, the basic requirement to have at least four jets and at least one b-jet is declared.

Since this search targets signatures with jets and missing transverse energy, the only possible triggers to use are jet or E_T^{miss} triggers. The lowest-p_T threshold unprescaled multi-jet triggers require the presence of either three jets with $p_T > 175$ GeV or four jets with $p_T > 85$ GeV each [24]. This requirement would significantly reduce the contributions from potential top squark pair production (cf. Fig. 11.6). Thus, the lowest unprescaled E_T^{miss} triggers are used. Thereby, the E_T^{miss} is calculated directly from the negative of the transverse momentum vector sum of all jets reconstructed by the jet trigger algorithm. Throughout the 2015 and 2016 data taking, due to the increasing instantaneous luminosity, in total four E_T^{miss} triggers were declared the lowest unprescaled ones. To estimate the trigger efficiency

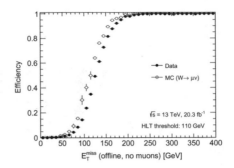

Fig. 11.5 E_T^{miss} trigger efficiency curve with respect to the E_T^{miss} reconstructed offline without muon corrections for all events passing a $W \rightarrow \mu\nu$ selection. Details on the specific selection as well as all efficiency curves of the trigger objects used throughout the full data-taking period can be found in Appendix B.1

in both measured and simulated data, events with a single muon with $p_T > 20\,\text{GeV}$ which triggered the TDAQ in addition to four jets and a b-jet are considered. The trigger efficiency is measured as the number of events which fired the respective trigger divided by all events containing four jets and one b=jet recorded due to the presence of a single muon trigger. The trigger efficiency curve for the trigger with the highest missing transverse energy threshold used during the 2015 and 2016 data taking period (cf. Fig. 11.5) shows that requiring $E_T^{miss} > 250\,\text{GeV}$ results in a constant trigger efficiency above 99.5%.

Figure 11.6 presents the relative distributions of transverse momenta of the four highest-p_T jets for the same selection of top squark signal processes shown in Fig. 11.4. More than 93% of the events have at least one jet with $p_T > 80\,\text{GeV}$ independently of the top squark and neutralino masses (cf. Fig. 11.6a). More than 90% of the second highest-p_T jets also satisfy $p_T > 80\,\text{GeV}$ (cf. Fig. 11.6b) and more than 98% of the third highest-p_T jets have a transverse momentum $p_T > 40\,\text{GeV}$ (cf. Fig. 11.6c). Figure 11.6a shows that for the fourth highest-p_T jets, at least 90% of the events satisfy $p_T > 40\,\text{GeV}$ apart from the signal process with a neutralino mass of 300 GeV where only 85% of the events contain at least four jets with $p_T > 40\,\text{GeV}$. However, since stronger requirements on the jet transverse momenta improve the SM background rejection, the p_T-requirement of the four jets leading in p_T is set to $(80, 80, 40, 40)\,\text{GeV}$.

Without applying any event selection, multi-jet production is by far the dominating process at the LHC. Thus, in final states with jets and missing transverse energy, multi-jet production needs to be suppressed by dedicated requirements in addition to the basic selection mentioned above. A priori, one would expect that strong requirements on the missing transverse energy, such as $E_T^{miss} > 250\,\text{GeV}$ are sufficient to suppress multi-jet production, since in case only SM processes are involved, neutrinos occurring in hadronisation processes are the only physical origin of missing transverse energy in multi-jet events which is expected to be small. However, the main origin of multi-jet processes remaining in selections with high missing transverse energy is

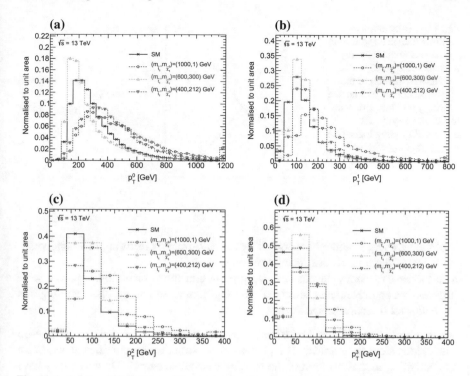

Fig. 11.6 Relative distributions of the jet transverse momenta p_T for the first four jets with the highest p_T after applying a lepton veto, $H_T > 150$ GeV, as well as at least four jets and one b-jet for the same selection of top squark signal processes shown in Fig. 11.4

the energy mis-measurement of multiple jets leading to an artificial imbalance of the transverse momenta. This leads to a degradation of the p_T-resolution of jets, It can be quantified in a broadening of the so-called jet response $R_{jet} = p_T^{reco} / p_T^{truth}$ where p_T^{reco} (p_T^{truth}) denotes the transverse momentum of the reconstructed jet (propagating parton). There are numerous sources of broadening of the jet response:

- No hadronic calorimeter can perform a perfect measurement due to the limitation by its granularity of the calorimeter cells resulting in a broad energy resolution.
- Since a jet is a particle shower with many constituents, it is possible that not all of the constituents are correctly picked up by anti-k_t algorithm which can cause a smaller total reconstructed energy of the jet.
- Depending on the direction of the jet, there may be significant amounts of dead material before the calorimeters such as service systems, support structure and damaged or inactive parts of the detector. If a jet cone partially overlaps with these regions, the jet will be reconstructed, but a smaller total reconstructed energy can be the consequence.
- Very highly energetic jets can punch through into the muon system when they are not fully absorbed in the hadronic calorimeter. In this case, a significant fraction

Fig. 11.7 Illustration of a multi-jet final state with four jets and no missing energy in the transverse plane (left). An underestimation of the transverse energy of a jet due to energy mis-measurements results in a significant amount of missing transverse energy aligning with the mis-measured jet (right)

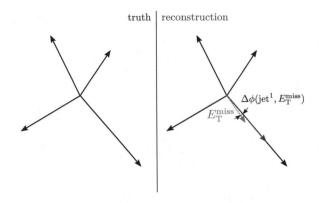

of the jet energy cannot be recorded by the calorimeters resulting in an underestimation of the total reconstructed energy.

- In the case of heavy flavour decays, in particular those involving b-quarks, neutrinos can be present and cause real missing transverse energy, also resulting in a smaller total reconstructed energy.

Other sources for additional missing transverse energy are the presence of pile-up jets, inconsistent the removal of lepton-jet reconstruction ambiguities as well as a mismodelling in the soft-term of the missing transverse energy. However, they have a minor impact on fake missing transverse energy than the effects of the energy mis-measurement of multiple jets.

Figure 11.7 illustrates the consequences of the energy mis-measurement of a jet in the event. Missing transverse energy is reconstructed aligning with the jet that was mis-measured and thus, multi-jet processes remain even after requiring $E_T^{miss} > 250\,\text{GeV}$. In order to suppress multi-jet contributions the minimum angular distance between the two jets leading in p_T and the missing energy in the transverse plane, $\left|\Delta\phi\left(\text{jet}^{0,1}, E_T^{miss}\right)\right|$, is calculated and shown in Fig. 11.8a for simulated multi-jet events and the selection of top squark signal processes shown in Fig. 11.4. Multi-jet processes tend to have one highly energetic jet which is very close to the missing energy in the transverse plane, since the transverse missing energy is a result of energy mis-measurements. For the top squark pair production processes, $\left|\Delta\phi\left(\text{jet}^{0,1}, E_T^{miss}\right)\right|$ is rather flat, since there, real missing transverse energy arising from the neutralinos is present. Merely, the simulation using a top squark mass of $400\,\text{GeV}$ and a neutralino mass of $212\,\text{GeV}$, which results in only slightly boosted neutralinos and thus, small missing transverse energy, also trends to have a $\left|\Delta\phi\left(\text{jet}^{0,1}, E_T^{miss}\right)\right|$ closer to 0 or π, but much less than the multi-jet processes. Figure 11.8b also shows the minimum angular distance, but using three instead of two jets leading in p_T, $\left|\Delta\phi\left(\text{jet}^{0,1,2}, E_T^{miss}\right)\right|$, which is smaller than 0.4 for more than 90% of the multi-jet events, but also more than 20% of the signal events with $m_{\tilde{t}_1} = 400\,\text{GeV}$ and $m_{\tilde{\chi}_1^0} = 212$ GeV. Thus, a requirement of $\left|\Delta\phi\left(\text{jet}^{0,1}, E_T^{miss}\right)\right| > 0.4$ is made for the basic selection of the search.

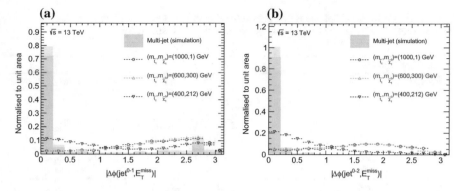

Fig. 11.8 Relative distributions of the minimum $\Delta\phi$ between the two (**a**) or three (**b**) jets leading in p_T, respectively, and the missing transverse energy for simulated multi-jets events and the same selection of top squark signal processes shown in Fig. 11.4. A selection vetoing leptons and requiring at least four jets and one b-jet is applied. The yellow bands show the quadratic sum of the statistical and experimental systematic uncertainties

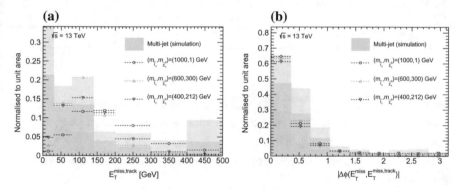

Fig. 11.9 Relative distributions of $E_T^{\text{miss,track}}$ and $\left|\Delta\phi\left(E_T^{\text{miss}}, E_T^{\text{miss,track}}\right)\right|$ for simulated multi-jets events and the same selection of top squark signal processes shown in Fig. 11.4. A selection vetoing leptons and requiring at least four jets, one b-jet and $\left|\Delta\phi\left(\text{jet}^{0,1}, E_T^{\text{miss}}\right)\right| > 0.4$ is applied. The yellow bands show the quadratic sum of the statistical and experimental systematic uncertainties

Another variable to additionally suppress multi-jet contributions, is the track-based missing transverse energy $E_T^{\text{miss,track}}$ shown in Fig. 11.9a after requiring $\left|\Delta\phi\left(\text{jet}^{0,1}, E_T^{\text{miss}}\right)\right| > 0.4$. Since $E_T^{\text{miss,track}}$ is calculated from ID tracks only, mis-measurements of energy deposits in the calorimeters do not have any effect. The requirement $E_T^{\text{miss,track}} > 30\,\text{GeV}$ rejects more than 21% of the multi-jet contributions while suppressing less than 5% of the signal processes. However, the size of the uncertainty band already indicates the large statistical and systematic uncertainties of the multi-jet simulation. Looking at the angular difference between the missing transverse energy and its track-based equivalent, $\left|\Delta\phi\left(E_T^{\text{miss}}, E_T^{\text{miss,track}}\right)\right|$, after requiring $\left|\Delta\phi\left(\text{jet}^{0,1}, E_T^{\text{miss}}\right)\right| > 0.4$ (cf. Fig. 11.9b), less than 10% of the signal processes are

Table 11.1 Preselection applied for all SRs defined in the search for fully hadronic decaying top squark pairs. $\mathscr{L}$ [10^{34} cm^{-2}s^{-1}] is denoting the instantaneous luminosity

	Requirement
Number of leptons	$N_\ell = 0$
Trigger	E_T^{miss}, HLT threshold: 70 GeV (2015)
	E_T^{miss}, HLT threshold: 90 GeV (2016, $\mathscr{L} \leq 1.02$)
	E_T^{miss}, HLT threshold: 100 GeV (2016, $1.02 < \mathscr{L} \leq 1.13$)
	E_T^{miss}, HLT threshold: 110 GeV (2016, $\mathscr{L} > 1.13$)
Missing transverse energy	$E_T^{miss} > 250$ GeV
Number of jets	$N_{jets} \geq 4$
Jet transverse momenta (sorted by p_T)	>80 GeV, >80 GeV, >40 GeV, >40 GeV
Number of b-jets	$N_{b-jets} \geq 1$
$\left| \Delta\phi \left(jet^{0,1}, E_T^{miss} \right) \right|$	>0.4
Track-based missing transverse energy $E_T^{miss,track}$	> 30 GeV
$\left| \Delta\phi \left(E_T^{miss}, E_T^{miss,track} \right) \right|$	$< \frac{\pi}{3}$

satisfying $\left| \Delta\phi \left(E_T^{miss}, E_T^{miss,track} \right) \right| > \frac{\pi}{3}$, while some additional multi-jet contributions can be suppressed. Thus, as a third requirement against multi-jet processes, $\left| \Delta\phi \left(E_T^{miss}, E_T^{miss,track} \right) \right| < \frac{\pi}{3}$ is made.

The resulting preselection used for all SRs defined in this search is summarised in Table 11.1.

11.3   Signal Region Definitions

Applying the basic selection summarised in Table 11.1 on the 36.1 fb^{-1} of collision data recorded in 2015 and 2016 as well as on the simulated SM processes introduced in Chap. 10, top quark pair production is the dominating SM contribution (cf. Fig. 11.10a). The major part of the $t\bar{t}$ contribution arises from top quark pairs where one top quark decays via a hadronic W boson decay and the other via a leptonic W boson decay. This is commonly referred to as semi-leptonic $t\bar{t}$ decay. The fully-hadronic $t\bar{t}$ decay is strongly suppressed by the $E_T^{miss} > 250$ GeV requirement since in this decay chain, no neutrinos are present and a large missing transverse energy can only be caused by enormous jet energy mis-measurements which is very unlikely. In order to suppress contributions from semi-leptonic $t\bar{t}$ decays, the transverse mass m_T of two particles, defined as

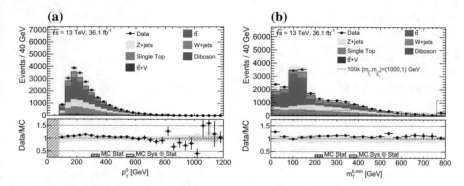

Fig. 11.10 Distributions of the transverse momentum of the highest-p_T jet, p_T^0 (a), and $m_T^{b,\min}$ (b) after the common preselection summarised in Table 11.1. The stacked histograms show the SM prediction before being normalised using normalisation factors derived from a simultaneous fit (detailed in Sect. 11.6) to all dominant SM backgrounds. The lower panels show the ratio of data events to the total SM prediction. The rightmost bin includes overflow events. The yellow bands show the quadratic sum of the statistical and experimental systematic uncertainties

$$m_T^2 = \left(E_{T,1} + E_{T,2}\right)^2 - \left(\mathbf{p}_{T,1} + \mathbf{p}_{T,2}\right)^2 \tag{11.1}$$

$$= m_1^2 + m_2^2 + 2\left(E_{T,1}E_{T,2} - \mathbf{p}_{T,1} \cdot \mathbf{p}_{T,2}\right) \quad , \tag{11.2}$$

is introduced. For the decay of a single particle of mass m into a visible and an invisible one, the distribution of m_T possesses an end-point at m [25]. In the approximation $m_1 = m_2 = 0$, Eq. (11.1) transforms into

$$m_T^2 = 2 \cdot |\mathbf{p}_{T,1}||\mathbf{p}_{T,2}| \left(1 - \cos(\Delta\phi_{12})\right) \quad , \tag{11.3}$$

with $\Delta\phi_{12}$ being the angular distance between the two particles in the transverse plane. Thus, the transverse mass of the b-jet closest in the azimuthal angle to the missing transverse momentum $\mathbf{p}_T^{\mathrm{miss}}$ and the missing transverse momentum, hence-forward called $m_T^{b,\min} = m_T(b_{\min\phi}, E_T^{\mathrm{miss}})$, has its endpoint at the top quark mass. The distribution of $m_T^{b,\min}$ obtained after the basic selection summarised in Table 11.1 is shown in Fig. 11.10b. Requiring $m_T^{b,\min} > 200$ GeV suppresses almost entirely the $t\bar{t}$ contribution while keeping most of the top squark signal contributions depicted by the dashed pink line.

Looking at the top squark-neutralino mass plane (cf. Fig. 11.11), the phase space can be categorised into various regions differing in the kinematics of the decay products. In order to obtain the highest possible sensitivity for each of the regions, specific additional requirements are made resulting in the definition of one (or more) SRs for each region of the phase space. Thereby, one categorises into:

- The region with high top squark masses and low neutralino masses (e.g. $m_{\tilde{t}_1} = 1000$ GeV and $m_{\tilde{\chi}_1^0} = 1$ GeV) where the high mass difference between top squark and top quark as well as the low neutralino mass result in a large Lorentz boost

Fig. 11.11 Schematic drawing of the top squark-neutralino mass plane including the parts of the phase space covered by the SRs targeting $\tilde{t}_1 \rightarrow t + \tilde{\chi}_1^0$ decays

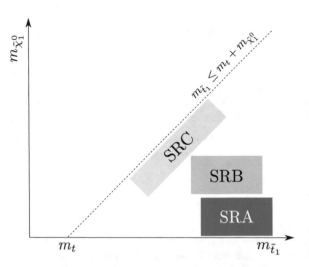

of the top squark decay products in the laboratory frame. Therefore, this region of phase space is also referred to as the boosted regime, in the following called SRA.
- The region with high top squark masses but intermediate neutralino masses (e.g. $m_{\tilde{t}_1} = 600\,\text{GeV}$ and $m_{\tilde{\chi}_1^0} = 300\,\text{GeV}$), in the following called SRB.
- The region with a mass difference between top squark and neutralino close to the top quark mass (e.g. $m_{\tilde{t}_1} = 400\,\text{GeV}$ and $m_{\tilde{\chi}_1^0} = 212\,\text{GeV}$) which is also referred to as the compressed scenario, in the following called SRC.
- Top squark decays into a bottom quark and a chargino which subsequently decays into a W boson and a neutralino (e.g. $m_{\tilde{t}_1} = 400\,\text{GeV}$, $m_{\tilde{\chi}_1^\pm} = 100\,\text{GeV}$ and $m_{\tilde{\chi}_1^0} = 50\,\text{GeV}$), in the following called SRD.

High Top Squark Masses

Usually, a fully-hadronic decay of a top quark pair results in six distinct $R = 0.4$ jets, one b-jet and two jets from the hadronic W boson decay for each top quark. However, for high top squark masses, the $\tilde{t}_1 \rightarrow t\tilde{\chi}_1^0$ decay can result in boosted top quarks which potentially leads to a distance between the top quark decay products smaller than $2R$ in the transverse plane. Thus, the $R = 0.4$ jets are reclustered using the anti-k_t algorithm with radius parameters of $R = 1.2$ and $R = 0.8$, respectively. Reclustering the $R = 0.4$ jets to larger objects has the advantage that the jet calibration and its uncertainties do not need to be re-estimated which would be the case if the reclustered jets were rebuild from the topological energy clusters. The masses of the two reclustered jets leading in p_T are shown in Fig. 11.12 for simulated top squark pair production with $m_{\tilde{t}_1} = 1000\,\text{GeV}$ and $m_{\tilde{\chi}_1^0} = 1\,\text{GeV}$. Looking at the reclustering with a radius parameter of $R = 1.2$ (cf. Fig. 11.12a), top squark events can be categorised based on the reclustered jet masses ordered in p_T. The category with both the first and second jet mass leading in p_T, $m_{\text{jet},R=1.2}^0$ and $m_{\text{jet},R=1.2}^1$, respectively, being around the top quark mass is labelled TT and defined by $m_{\text{jet},R=1.2}^0 > 120\,\text{GeV}$

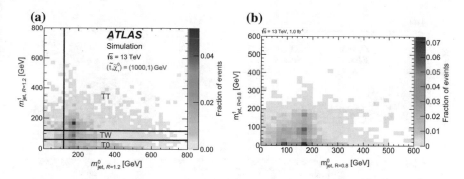

Fig. 11.12 Illustration of the SR categories TT, TW and T0 based on the $R = 1.2$ (**a**) and $R = 0.8$ (**b**) reclustered jet masses for simulated top squark pair production with $m_{\tilde{t}_1} = 1000\,\mathrm{GeV}$ and $m_{\tilde{\chi}_1^0} = 1\,\mathrm{GeV}$ after the preselection requirements (cf. Table 11.1). The black lines in (**a**) represent the requirements on the reclustered jet masses [22]

and $m^1_{\mathrm{jet},\,R=1.2} > 120\,\mathrm{GeV}$. If the reclustered jet mass of the second highest p_T jet is around the W boson mass ($60\,\mathrm{GeV} < m^1_{\mathrm{jet},\,R=1.2} < 120\,\mathrm{GeV}$), it is categorised as TW, while when it is below ($m^1_{\mathrm{jet},\,R=1.2} < 60\,\mathrm{GeV}$), the event is labelled as T0. Reclustering the $R = 0.4$ jets with a radius parameter of $R = 0.8$ (cf. Fig. 11.12b), the reclustered jet masses are less pronounced which hints to the fact, that the boost of the top is not sufficiently high enough such that all of its decay products fall into a cone of $R = 0.8$. Therefore, the categories TT, TW and T0 based on the $R = 1.2$ reclustered jets are exploited in the SR definitions targeting high top squark masses.

In order to increase the sensitivity to high top squark masses but low neutralino masses, an optimisation based on a statistical hypothesis test using simulated data scaled to an integrated luminosity of $36.1\,\mathrm{fb}^{-1}$ is performed in order to reflect the size of the dataset taken in 2015 and 2016. For the simulation of top squark pair production, the mass parameters $m_{\tilde{t}_1} = 1000\,\mathrm{GeV}$ and $m_{\tilde{\chi}_1^0} = 1\,\mathrm{GeV}$ are used. Assuming that only SM processes are present, the p-value for observing both SM and top squark signal events is calculated using a one-sided test statistics [26] assuming an uncertainty of 30% on the expected SM yield which is justified by previous searches for top squarks in the fully hadronic decay channel [1]. Subsequently, the p-value is translated into the number of standard deviations of a Gaussian distribution, the so-called *significance*, by the inverse of the Gaussian error function using the ROOT software toolkit [27]. Thereby, the significance is only computed in case the statistical uncertainty of both simulated signal and SM contributions is less then 50% in order to avoid unphysical results.

Figure 11.13a shows the simulated distribution of E_T^{miss} after applying the basic requirements summarised in Table 11.1 with the expected significance of the signal scenario with $m_{\tilde{t}_1} = 1000\,\mathrm{GeV}$ and $m_{\tilde{\chi}_1^0} = 1\,\mathrm{GeV}$ depending on the cut value on E_T^{miss} in the bottom panel. Requiring $E_T^{\mathrm{miss}} > 400\,\mathrm{GeV}$ does not lead to any loss of expected significance and thus, is part of the basic requirements for SRA. The distribution of the stransverse mass m_{T2} [28, 29], an additional variable used to model

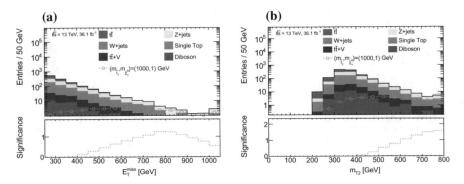

Fig. 11.13 Simulated distributions of E_T^{miss} (**a**) and m_{T2} (**b**) with the the expected significance of the signal scenario with $m_{\tilde{t}_1} = 1000$ GeV and $m_{\tilde{\chi}_1^0} = 1$ GeV depending on a potential cut value on the corresponding variable in the bottom panel. In addition to the basic requirements summarised in Table 11.1, at least two b-jets and $m_T^{b,min} > 200$ GeV is required

the fully-hadronic top quark decay is depicted in Fig. 11.13b. The calculation of m_{T2} requires the presence of two b-jets, thus, two b-jets are required for all high mass top squark SRs. Only the two b-jets with the highest MV2 discriminant are considered as b-jets for the calculation of m_{T2}. Consecutively, all non-b-jets are combined in pairs and their invariant mass is calculated as W boson candidate mass m_{W-cand}. Subsequently, two W boson candidates are paired with the two b-jets in order to form a top quark candidate of mass m_{t-cand}. Minimising $\chi^2 = (m_{t-cand} - m_t)^2/m_t + (m_{W-cand} - m_W)^2/m_W$ with m_t and m_W being the real top quark and W boson masses, respectively, two top quark candidates are identified. Finally, the stransverse mass m_{T2} is calculated from the top quark candidates and the missing transverse energy setting the top quark candidate masses to 173.2 GeV and the invisible particle mass to 0 GeV.

As for the requirement on E_T^{miss}, requiring $m_{T2} > 400$ GeV does not lead to any loss of expected significance (cf. Fig. 11.13b) and thus, is part of the basic requirements for SRA.

The simulated distribution of the distance between the two b-jets with the highest MV2 discriminant, $\Delta R(b, b)$, for the TT category is depicted in Fig. 11.14 after applying all requirements mentioned in Fig. 11.13 as well as the requirements on E_T^{miss} and m_{T2}. The expected significance reaches its maximum for a $\Delta R(b, b) > 1$ requirement which is applied for the TT category of SRA. Requiring $\Delta R(b, b) > 1$ provides additional discrimination against SM processes which involve a gluon splitting into a $b\bar{b}$ pair, such as $V +$ (heavy-flavour) jets.

The simulated distributions of m_{T2} in the TW and T0 categories are shown in Fig. 11.15a and 11.15b, respectively. Requirements on $m_{T2} > 400$ GeV and $m_{T2} > 500$ GeV can be applied without loosing any expected significance in the TW and T0 categories.

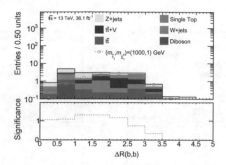

Fig. 11.14 Simulated distribution of $\Delta R\,(b, b)$ after requiring $E_T^{\text{miss}} > 400\,\text{GeV}$ and $m_{T2} > 400\,\text{GeV}$ on top of the selection mentioned in Fig. 11.13 for the TT category of SRA. The bottom panel shows the expected significance for the signal scenario with $m_{\tilde{t}_1} = 1000\,\text{GeV}$ and $m_{\tilde{\chi}_1^0} = 1\,\text{GeV}$ depending on a potential cut value on the corresponding variable

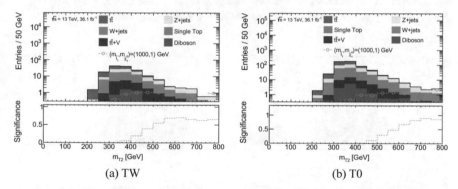

(a) TW
(b) T0

Fig. 11.15 Simulated distributions of m_{T2} after applying the basic preselection summarised in Table 11.1 as well as requiring at least two b-jets and $m_T^{b,\min} > 200\,\text{GeV}$ in the TW (**a**) and T0 (**b**) categories, respectively. The bottom panel shows the expected significance for the signal scenario with $m_{\tilde{t}_1} = 1000\,\text{GeV}$ and $m_{\tilde{\chi}_1^0} = 1\,\text{GeV}$ depending on a potential cut value on the corresponding variable

After additionally requiring of $m_{T2} > 400\,\text{GeV}$ in the TW category, the simulated distributions of E_T^{miss} and the mass of the $R = 0.8$ reclustered jet leading in p_T, $m_{\text{jet},R=0.8}^0$, are shown in Fig. 11.16. Although the expected significance of the $m_{\tilde{t}_1} = 1000\,\text{GeV}$ and $m_{\tilde{\chi}_1^0} = 1\,\text{GeV}$ signal scenario reaches its maximum at a $E_T^{\text{miss}} > 750\,\text{GeV}$ requirement, looking at the scenario with $m_{\tilde{t}_1} = 800\,\text{GeV}$ and $m_{\tilde{\chi}_1^0} = 1\,\text{GeV}$ shows that a $E_T^{\text{miss}} > 750\,\text{GeV}$ requirement would significantly reduce the expected significance for a signal scenario with a smaller top squark mass. Thus, a $E_T^{\text{miss}} > 500\,\text{GeV}$ requirement is made for the TW category of SRA. The same argument leads to requiring $m_{\text{jet},R=0.8}^0 > 60\,\text{GeV}$ (cf. Fig. 11.16b).

Analogously, the simulated distributions of E_T^{miss} and $m_{\text{jet},R=0.8}^0$ are shown after requiring $m_{T2} > 500\,\text{GeV}$ for the T0 category of SRA in Fig. 11.17. Following the arguments from the TW category, $E_T^{\text{miss}} > 550\,\text{GeV}$ and $m_{\text{jet},R=0.8}^0 > 60\,\text{GeV}$ are

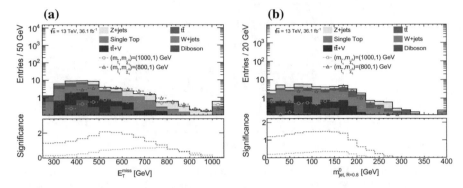

Fig. 11.16 Simulated distributions E_T^{miss} (**a**) and $m_{jet, R=0.8}^0$ (**b**) after applying $m_{T2} > 400$ GeV and the selection mentioned in Fig. 11.15a in the TW category of SRA. The bottom panel shows the expected significance for the signal scenarios with $(m_{\tilde{t}_1}, m_{\tilde{\chi}_1^0}) = (1000, 1)$ GeV and $(m_{\tilde{t}_1}, m_{\tilde{\chi}_1^0}) = (800, 1)$ GeV depending on a potential cut value on the corresponding variable

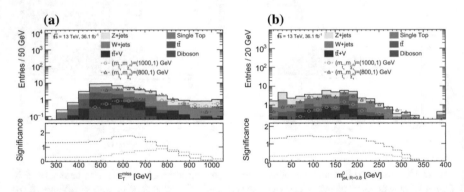

Fig. 11.17 Simulated distributions E_T^{miss} (**a**) and $m_{jet, R=0.8}^0$ (**b**) after applying $m_{T2} > 500$ GeV and the selection mentioned in Fig. 11.15b in the T0 category of SRA. The bottom panel shows the expected significance for the signal scenarios with $(m_{\tilde{t}_1}, m_{\tilde{\chi}_1^0}) = (1000, 1)$ GeV and $(m_{\tilde{t}_1}, m_{\tilde{\chi}_1^0}) = (800, 1)$ GeV depending on a potential cut value on the corresponding variable

added to the requirements of the T0 category of SRA. Since both the TW and T0 category of SRA contain the $m_{jet, R=0.8}^0 > 60$ GeV requirement and no loss of expected sensitivity is observed when also adding the requirement to the TT category (cf. Appendix B.3), $m_{jet, R=0.8}^0 > 60$ GeV is a common requirement in SRA. All requirements for all categories of SRA are summarised in Table 11.2.

For high top squark masses but intermediate neutralino masses, the same optimisation is performed using the mass parameters $(m_{\tilde{t}_1}, m_{\tilde{\chi}_1^0}) = (700, 400)$ GeV and $(m_{\tilde{t}_1}, m_{\tilde{\chi}_1^0}) = (600, 300)$ GeV, respectively. After requiring the presence of at least two b-jets and $m_T^{b,min} > 200$ GeV in addition to the basic requirements summarised in Table 11.1, the distributions of $\Delta R(b, b)$ and the transverse mass of the b-jet farthest from the direction of $\mathbf{p}_T^{miss}$ and the missing transverse energy, $m_T^{b,max}$, are

Table 11.2 Signal region selection targeting high top squark masses in addition to the requirements presented in Table 11.1

Signal region	Variable	TT	TW	T0
	$m^0_{jet, R=1.2}$		>120 GeV	
	$m^1_{jet, R=1.2}$	>120 GeV	60–120 GeV	<60 GeV
	$m^{b,min}_T$		>200 GeV	
	b-tagged jets		≥ 2	
	τ-veto		Yes	
	$\left|\Delta\phi\left(\mathrm{jet}^{0,1,2}, E_T^{miss}\right)\right|$		>0.4	
A	$m^0_{jet, R=0.8}$		>60 GeV	
	$\Delta R\,(b, b)$	>1	–	
	m_{T2}	>400 GeV	>400 GeV	>500 GeV
	E_T^{miss}	>400 GeV	>500 GeV	>550 GeV
B	$m^{b,max}_T$		>200 GeV	
	$\Delta R\,(b, b)$		>1.2	

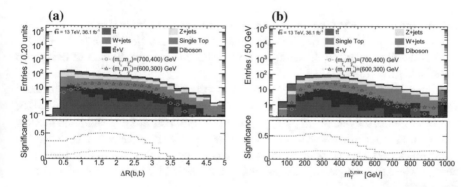

Fig. 11.18 Simulated distributions $\Delta R\,(b, b)$ (**a**) and $m^{b,max}_T$ (**b**) after requiring the presence of at least two b-jets and $m^{b,min}_T > 200$ GeV in addition to the basic requirements summarised in Table 11.1. The bottom panel shows the expected significance for the signal scenarios with $(m_{\tilde{t}_1}, m_{\tilde{\chi}^0_1}) = (700, 400)$ GeV and $(m_{\tilde{t}_1}, m_{\tilde{\chi}^0_1}) = (600, 300)$ GeV depending on a potential cut value on the corresponding variable. For (b), $\Delta R\,(b, b) > 1.2$ is required

shown in Fig. 11.18. The expected significances reach their maxima for $\Delta R\,(b, b) > 1.2$ (cf. Fig. 11.18a) and, after applying this requirement, for $m^{b,max}_T > 200$ GeV.

In order to suppress hadronic τ decays which are not eliminated by the lepton veto, events that contain a non-b-jet within $|\eta| < 2.5$ with fewer than four associated charged-particle tracks with $p_T > 500$ MeV are discarded in case the angular separation between the jet and the $\mathbf{p}_T^{miss}$ in the transverse plane is less than $\pi/5$. The systematic uncertainties for the τ-veto were shown to be negligible [1]. Additionally, for

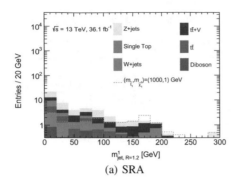 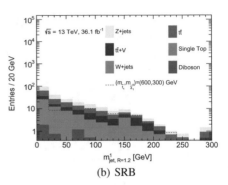

(a) SRA   (b) SRB

Fig. 11.19 Distributions of $m^1_{\text{jet}, R=1.2}$ in SRA (**a**) and SRB (**b**) after applying all requirements listed in Table 11.2 but the one on $m^1_{\text{jet}, R=1.2}$ scaled to an integrated luminosity of 36.1 fb^{-1} without drawing statistical or systematic uncertainty bands. In addition to the SM processes, also simulated top squark pair production with $(m_{\tilde{t}_1}, m_{\tilde{\chi}^0_1}) = (1000, 1)$ GeV (**a**) and $(m_{\tilde{t}_1}, m_{\tilde{\chi}^0_1}) = (600, 300)$ GeV (**b**) are shown, respectively. The rightmost bin includes overflow events

entirely ensuring the suppression of multi-jet processes (discussed in Sect. 11.4), the $\left|\Delta\phi\left(\text{jet}^{0,1}, E_T^{\text{miss}}\right)\right| > 0.4$ requirement is tightened to $\left|\Delta\phi\left(\text{jet}^{0,1,2}, E_T^{\text{miss}}\right)\right| > 0.4$.

A full summary of all SR definitions covering the pair production of heavy top squarks is given in Table 11.2.

Figure 11.19a, b are showing the mass distribution of the $R = 1.2$ reclustered jet with the second largest p_T, $m^1_{\text{jet}, R=1.2}$ after applying all cuts of SRA and SRB, respectively but the one on $m^1_{\text{jet}, R=1.2}$. The most dominant SM contributions in SRA (SRB) are arising from Z+jets and $t\bar{t} + Z$ ($t\bar{t}$) production, respectively. Thereby, Z boson decays into pairs of neutrinos are creating additional missing transverse energy. The expected yields in all SRs defined in Table 11.2 for a dataset of 36.1 fb^{-1} are shown in Table 11.3.

Compressed Top Squark Decay Scenarios

Models in which the mass difference between top squark and neutralino is around the top quark mass, $m_{\tilde{t}_1} - m_{\tilde{\chi}^0_1} \gtrsim m_t$, are of special interest since after LHC Run 1, there was a significant part of the phase space not excluded in the top squark-neutralino mass plane (cf. Fig. 11.2). The main experimental challenge in this region of the phase space is to find a way to discriminate those scenarios from SM top quark pair production, because in case of $m_{\tilde{t}_1} - m_{\tilde{\chi}^0_1} \gtrsim m_t$, almost no missing transverse energy is created which makes the process hard to distinguish from non-resonant $t\bar{t}$ production. Since the top squarks are produced back to back, the little additional energy available for the neutralinos does not allow any other than the neutralino and the top quark moving approximately collinearly with the original top squark boost axis. Hence, the two momenta of the neutralinos will roughly cancel resulting in a small missing transverse energy below the plateau of the trigger efficiency curve. The main idea for these so-called compressed scenarios is that one requires the presence of highly energetic initial-state radiation (ISR) in form of one additional high energetic jet which boosts the top squark pair system in the transverse plane (cf. Fig. 11.20a).

Table 11.3 Expected number of simulated events in the signal regions targeting high top squark masses scaled to an integrated luminosity of 36.1 fb^{-1}. For the signal scenarios used for the optimisation of the SRs, the expected significance is given. For the SM processes, the statistical and the experimental systematic uncertainties are shown

	SRA-TT	SRA-TW	SRA-T0	SRB-TT	SRB-TW	SRB-T0
$(m_{\tilde{t}_1}, m_{\tilde{\chi}_1^0}) = (1000, 1)$ GeV	$8.22 \pm 0.37\ (1.8\sigma)$	$4.61 \pm 0.27\ (0.9\sigma)$	$6.43 \pm 0.33\ (0.7\sigma)$	$10.07 \pm 0.40\ (0.7\sigma)$	$5.41 \pm 0.30\ (0.1\sigma)$	$5.63 \pm 0.30\ (0.0\sigma)$
$(m_{\tilde{t}_1}, m_{\tilde{\chi}_1^0}) = (700, 400)$ GeV	$1.90 \pm 0.15\ (0.3\sigma)$	$1.11 \pm 0.14\ (0.1\sigma)$	$0.35 \pm 0.06\ (0.0\sigma)$	$10.02 \pm 0.32\ (0.7\sigma)$	$13.34 \pm 0.40\ (0.6\sigma)$	$21.51 \pm 0.50\ (0.2\sigma)$
$(m_{\tilde{t}_1}, m_{\tilde{\chi}_1^0}) = (600, 300)$ GeV	$2.76 \pm 0.41\ (0.5\sigma)$	$1.51 \pm 0.23\ (0.2\sigma)$	$0.61 \pm 0.15\ (0.0\sigma)$	$20.43 \pm 1.05\ (1.4\sigma)$	$25.98 \pm 1.13\ (1.3\sigma)$	$43.58 \pm 1.39\ (0.6\sigma)$
$t\bar{t}$	$0.60 \pm 0.10^{+0.174}_{-0.150}$	$0.45 \pm 0.12^{+0.120}_{-0.181}$	$1.45 \pm 0.31^{+0.470}_{-0.452}$	$6.10 \pm 0.59^{+1.107}_{-1.285}$	$12.81 \pm 1.03^{+2.516}_{-3.040}$	$47.23 \pm 2.04^{+8.558}_{-6.928}$
Z+jets	$2.15 \pm 0.27^{+0.640}_{-0.663}$	$4.20 \pm 0.50^{+0.826}_{-0.812}$	$8.63 \pm 0.65^{+1.455}_{-1.154}$	$7.72 \pm 0.66^{+1.234}_{-1.280}$	$14.41 \pm 1.07^{+3.227}_{-2.890}$	$53.99 \pm 1.81^{+8.685}_{-6.960}$
W+jets	$0.65 \pm 0.16^{+0.116}_{-0.082}$	$0.70 \pm 0.20^{+0.147}_{-0.432}$	$1.58 \pm 0.47^{+0.636}_{-0.679}$	$6.12 \pm 2.74^{+3.000}_{-3.081}$	$3.83 \pm 0.61^{+1.093}_{-0.970}$	$20.39 \pm 3.15^{+6.414}_{-6.959}$
Single Top	$1.03 \pm 0.55^{+0.421}_{-0.318}$	$0.60 \pm 0.15^{+0.176}_{-0.177}$	$2.52 \pm 1.02^{+1.258}_{-1.176}$	$3.59 \pm 0.64^{+0.618}_{-0.607}$	$5.16 \pm 0.46^{+0.803}_{-0.911}$	$22.53 \pm 1.60^{+2.079}_{-3.372}$
$t\bar{t} + V$	$2.46 \pm 0.28^{+0.277}_{-0.345}$	$1.43 \pm 0.21^{+0.282}_{-0.188}$	$2.02 \pm 0.21^{+0.222}_{-0.227}$	$7.25 \pm 0.50^{+0.902}_{-0.786}$	$8.37 \pm 0.51^{+1.071}_{-0.882}$	$15.92 \pm 0.62^{+1.841}_{-1.346}$
Diboson	$0.12 \pm 0.08^{+0.148}_{-0.129}$	$0.34 \pm 0.14^{+0.072}_{-0.054}$	$0.84 \pm 0.31^{+0.271}_{-0.270}$	$0.23 \pm 0.11^{+0.185}_{-0.149}$	$1.84 \pm 0.57 \pm 0.467$	$2.94 \pm 0.46^{+0.437}_{-0.491}$
Total SM	$7.01 \pm 0.70^{+1.045}_{-1.018}$	$7.72 \pm 0.62^{+1.186}_{-1.118}$	$17.05 \pm 1.38^{+3.521}_{-2.246}$	$31.01 \pm 2.99^{+4.500}_{-4.795}$	$46.43 \pm 1.84^{+8.046}_{-7.823}$	$163.00 \pm 4.53^{+24.198}_{-20.590}$

(a)   **(b)**

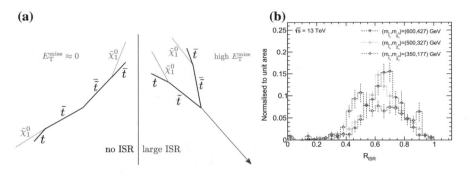

Fig. 11.20 Illustration of the impact of a high energetic ISR jet on the top squark pair system and its decay products (**a**) and correlation between the ratio of missing transverse energy and p_T of the ISR system and the mass ratio of neutralino and top squark (**b**)

In order to disentangle between the ISR and the top squark pair system, a recursive jigsaw reconstruction (RJR) technique [30, 31] is exploited. The RJR algorithm iteratively maximises the back to back momenta of all visible objects in the transverse plane in the reference frames of every step in the decay chain. The resulting hemispheres at each decay step allow to calculate the variables of interest in the corresponding reference frame. In the end, two frames are defined, the sparticle frame containing the top squarks and the ISR frame, representing all objects not coming from the sparticle decay. Since the basic preselection described in Table 11.1 does not explicitly require the presence of a highly energetic ISR jet, the RJR algorithm does not necessarily assign the jets correctly into the sparticle and ISR hemispheres. Thus a requirement on the transverse momentum of the ISR system, $p_T^{\text{ISR}} > 400$ GeV, is made, in order to obtain a high efficiency of the RJR. The ISR jet boosts the top squark pair system and consequently also the neutralino pair. The ratio between the missing transverse energy and the transverse momentum of the ISR system p_T^{ISR} which is proportional to the ratio of neutralino and top squark mass [32, 33],

$$R_{\text{ISR}} = \frac{E_T^{\text{miss}}}{p_T^{\text{ISR}}} \propto \frac{m_{\tilde{\chi}_1^0}}{m_{\tilde{t}_1}} \quad , \tag{11.4}$$

is depicted in Fig. 11.20b for three simulated scenarios of top squark pair production with $m_{\tilde{t}_1} - m_{\tilde{\chi}_1^0} \gtrsim m_t$. The distributions of R_{ISR} are normalised to unit area and perfectly peak at the neutralino-top squark mass ratio.

Since the RJR algorithm is defining the sparticle (S) and the ISR frame, instead of using the laboratory frame, the optimisation of the selection for the compressed scenarios is performed with the kinematic variables defined in the S and ISR frames, respectively. Simulated top squark pair production with the mass parameters $m_{\tilde{t}_1} = 450$ GeV, $m_{\tilde{\chi}_1^0} = 277$ GeV is used to calculate the expected significances.

The simulated distribution of transverse momentum of the fifth-leading jet in the sparticle system, $p_T^{4,S}$, is shown in Fig. 11.21a after applying the basic requirements summarised in Table 11.1 as well as $p_T^{\text{ISR}} > 400$ GeV. The maximum expected

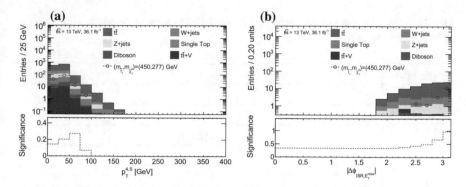

Fig. 11.21 Simulated distributions of $p_T^{4,S}$ (**a**) and $\Delta\phi_{\text{ISR},E_T^{\text{miss}}}$ (**b**) after applying the basic requirements summarised in Table 11.1 as well as $p_T^{\text{ISR}} > 400$ GeV. For (**b**), also $p_T^{4,S} > 50$ GeV is required. The bottom panel shows the expected significance for the signal scenario with $m_{\tilde{t}_1} = 450$ GeV and $m_{\tilde{\chi}_1^0} = 277$ GeV depending on a potential cut value on the corresponding variable

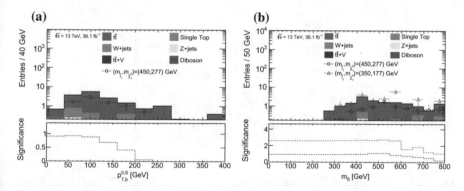

Fig. 11.22 Simulated distributions of $p_{T,b}^{0,S}$ (**a**) and m_S (**b**) after applying the basic requirements summarised in Table 11.1 as well as $p_T^{\text{ISR}} > 400$ GeV, $p_T^{4,S} > 50$ GeV and $|\Delta\phi_{\text{ISR},E_T^{\text{miss}}}| > 3$

sensitivity for the signal scenario with $m_{\tilde{t}_1} = 450$ GeV and $m_{\tilde{\chi}_1^0} = 277$ GeV is reached when requiring $p_T^{4,S} > 50$ GeV. Figure 11.21b shows the distribution of the angular distance between the ISR system and the E_T^{miss}, $\Delta\phi_{\text{ISR},E_T^{\text{miss}}}$, after applying additionally $p_T^{4,S} > 50$ GeV. Signal like events are accumulated at $|\Delta\phi_{\text{ISR},E_T^{\text{miss}}}| \sim \pi$, since the E_T^{miss} arises from the neutralinos which ideally are located back to back with respect to the ISR jet. A $|\Delta\phi_{\text{ISR},E_T^{\text{miss}}}| > 3$ requirement is added for SRC.

The simulated distribution of the transverse momentum of the b-jet in the sparticle frame leading in p_T, $p_{T,b}^{0,S}$, is depicted in Fig. 11.22a. Besides the basic requirements summarised in Table 11.1 and $p_T^{\text{ISR}} > 400$ GeV, also $p_T^{4,S} > 50$ GeV and $|\Delta\phi_{\text{ISR},E_T^{\text{miss}}}| > 3$ are required. The maximum of the expected significance can be reached by requiring $p_{T,b}^{0,S} > 40$ GeV. With this requirement applied, the an addi-

Table 11.4 Signal region selection targeting compressed decay scenarios in addition to the requirements presented in Table 11.1

Variable	SRC-1	SRC-2	SRC-3	SRC-4	SRC-5
b-tagged jets			≥ 1		
$N^S_{b-\text{jet}}$			≥ 1		
N^S_{jet}			≥ 5		
$p^{0,S}_{T,b}$			$> 40\,\text{GeV}$		
m_S			$> 300\,\text{GeV}$		
$\Delta\phi_{\text{ISR},E^{\text{miss}}_T}$			> 3.0		
p^{ISR}_T			$> 400\,\text{GeV}$		
$p^{4,S}_T$			$> 50\,\text{GeV}$		
R_{ISR}	0.30–0.40	0.40–0.50	0.50–0.60	0.60–0.70	0.70–0.80

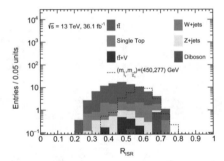

Fig. 11.23 Distribution of R_{ISR} after applying all requirements listed in Table 11.4 but the one on R_{ISR} scaled to an integrated luminosity of $36.1\,\text{fb}^{-1}$ without drawing statistical or systematic uncertainty bands. In addition to the SM processes, also simulated top squark pair production with $m_{\tilde{t}_1} = 450\,\text{GeV}$ and $m_{\tilde{\chi}^0_1} = 277\,\text{GeV}$ is shown. The rightmost bin includes overflow events

tional requirement on the transverse mass between the sparticle frame and the E^{miss}_T, m_S (cf. Fig. 11.22b) does not increase the expected significance for the signal scenario with $m_{\tilde{t}_1} = 450\,\text{GeV}$ and $m_{\tilde{\chi}^0_1} = 277\,\text{GeV}$ anymore. However, depending on the compressed signal scenario considered, a requirement tighter than $m_S > 300\,\text{GeV}$ would already lead to a loss of MC statistics, thus $m_S > 300\,\text{GeV}$ is chosen in order to preserve the statistics of the signal simulations.

Figure 11.20b shows that there is a correlation between the ratio of missing transverse energy and p_T of the ISR system, R_{ISR}, and the ratio $m_{\tilde{\chi}^0_1}/m_{\tilde{t}_1}$. So, the final SRs for top squark pair production with $m_{\tilde{t}_1} - m_{\tilde{\chi}^0_1} \gtrsim m_t$ are divided by the value of R_{ISR} (cf. Table 11.4).

Figure 11.23 shows the distribution of R_{ISR}, after applying all cuts of SRC except R_{ISR} itself. The most dominant SM contribution arises from $t\bar{t}$ production. The expected yields in all SRs defined in Table 11.4 for a dataset of $36.1\,\text{fb}^{-1}$ are shown in Table 11.5.

Table 11.5 Expected number of simulated events in the signal regions targeting compressed top squark decay scenarios scaled to an integrated luminosity of 36.1 fb^{-1}. For the signal scenarios used for the optimisation of the SRs, the expected significance is given. For the SM processes, the statistical and the experimental systematic uncertainties are shown

	SRC-1	SRC-2	SRC-3	SRC-4	SRC-5
$(m_{\tilde{t}_1}, m_{\tilde{\chi}_1^0}) = (250, 77)$ GeV	35.23 ± 4.18 (3.1σ)	17.40 ± 2.46 (1.1σ)	6.99 ± 1.61 (0.5σ)	1.58 ± 0.65 (0.1σ)	0.00 ± 0.00 (0.0σ)
$(m_{\tilde{t}_1}, m_{\tilde{\chi}_1^0}) = (300, 127)$ GeV	28.60 ± 3.89 (2.6σ)	67.31 ± 6.79 (3.7σ)	13.80 ± 2.39 (1.2σ)	0.74 ± 0.53 (0.0σ)	0.52 ± 0.36 (0.1σ)
$(m_{\tilde{t}_1}, m_{\tilde{\chi}_1^0}) = (350, 177)$ GeV	2.50 ± 0.80 (0.1σ)	30.54 ± 3.40 (1.9σ)	46.16 ± 5.05 (3.5σ)	7.45 ± 2.15 (1.3σ)	0.37 ± 0.37 (0.0σ)
$(m_{\tilde{t}_1}, m_{\tilde{\chi}_1^0}) = (450, 277)$ GeV	0.62 ± 0.22 (0.0σ)	4.15 ± 0.61 (0.1σ)	16.22 ± 1.23 (1.4σ)	12.19 ± 0.94 (2.1σ)	0.90 ± 0.24 (0.4σ)
$(m_{\tilde{t}_1}, m_{\tilde{\chi}_1^0}) = (600, 427)$ GeV	0.00 ± 0.00 (0.0σ)	0.47 ± 0.21 (0.0σ)	1.66 ± 0.63 (0.0σ)	4.61 ± 0.76 (0.8σ)	0.61 ± 0.21 (0.2σ)
$t\bar{t}$	$18.21 \pm 1.58^{+1.980}_{-2.259}$	$31.22 \pm 1.82^{+5.955}_{-5.069}$	$20.63 \pm 1.07^{+1.403}_{-3.081}$	$6.96 \pm 0.46^{+1.142}_{-1.008}$	$0.89 \pm 0.24^{+0.475}_{-0.449}$
Z+jets	$0.46 \pm 0.09^{+0.110}_{-0.083}$	$0.90 \pm 0.13^{+0.171}_{-0.202}$	$0.75 \pm 0.15^{+0.200}_{-0.181}$	$0.45 \pm 0.10^{+0.160}_{-0.139}$	$0.09 \pm 0.03 \pm 0.121$
W+jets	$0.64 \pm 0.13^{+0.278}_{-0.496}$	$1.51 \pm 0.31^{+0.417}_{-0.251}$	$1.51 \pm 0.37^{+0.428}_{-0.535}$	$1.53 \pm 0.41^{+0.411}_{-0.401}$	$0.17 \pm 0.09^{+0.089}_{-0.075}$
Single top	$1.50 \pm 0.51^{+1.286}_{-1.141}$	$1.02 \pm 0.19^{+1.133}_{-1.341}$	$1.05 \pm 0.41^{+0.343}_{-0.410}$	$0.62 \pm 0.17^{+0.169}_{-0.152}$	0.00 ± 0.00
$t\bar{t}+V$	$0.22 \pm 0.10^{+0.142}_{-0.106}$	$0.46 \pm 0.19^{+0.398}_{-0.168}$	$0.44 \pm 0.11^{+0.284}_{-0.214}$	$0.07 \pm 0.08^{+0.098}_{-0.103}$	$0.05 \pm 0.03^{+0.014}_{-0.007}$
Diboson	$0.43 \pm 0.29^{+0.256}_{-0.252}$	$0.24 \pm 0.21^{+0.050}_{-0.048}$	$0.40 \pm 0.30 \pm 0.161$	$0.23 \pm 0.16^{+0.162}_{-0.154}$	0.00 ± 0.00
Total SM	$21.46 \pm 1.69^{+2.160}_{-2.811}$	$35.35 \pm 1.88^{+7.063}_{-6.189}$	$24.78 \pm 1.26^{+1.797}_{-3.409}$	$9.85 \pm 0.67^{+1.425}_{-1.319}$	$1.12 \pm 0.36^{+0.497}_{-0.394}$

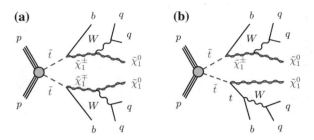

Fig. 11.24 Selection of schematic Feynman graphs for direct $\tilde{t}_1$ production and possible subsequent fully hadronic decays for $m_{\tilde{t}_1} > m_{\tilde{\chi}_1^\pm} > m_{\tilde{\chi}_1^0}$. The two-step decay can either be symmetric (**a**) or a mixed scenario (**b**) with one top squark decaying directly into a top quark and a neutralino. Note that all graphs result in the same final state

Top Squark Decays Involving Charginos

In case the supersymmetric model allows the lightest chargino $\tilde{\chi}_1^\pm$ to be lighter than the top squark, top squarks can decay into charginos and bottom quarks with a subsequent decay of the chargino into a W boson and the neutralino. Either one or both top squarks can undergo the decay involving the chargino (cf. Fig. 11.24). For top squark decays involving charginos, only two-step decays are considered and $m_{\tilde{\chi}_1^\pm} = 2 \cdot m_{\tilde{\chi}_1^0}$ is required, which is motivated by models with gauge unification at the GUT scale [34–36].

The optimisation of SRs for top squark decays involving charginos is performed using simulated signal events with $(m_{\tilde{t}_1}, m_{\tilde{\chi}_1^\pm}, m_{\tilde{\chi}_1^0}) = (400, 100, 50)\,\text{GeV}$ and $(m_{\tilde{t}_1}, m_{\tilde{\chi}_1^\pm}, m_{\tilde{\chi}_1^0}) = (700, 100, 50)\,\text{GeV}$, respectively. For light neutralinos, the assumption $m_{\tilde{\chi}_1^\pm} = 2 \cdot m_{\tilde{\chi}_1^0}$ causes a larger mass difference between top squark and chargino than between chargino and neutralino, $\Delta m(\tilde{t}_1, \tilde{\chi}_1^\pm) > \Delta m(\tilde{\chi}_1^\pm, \tilde{\chi}_1^0)$ which leads to the presence of b-jets with high transverse momenta. Figure 11.25a shows the simulated distribution of the scalar sum of transverse momenta of the two b-jets with the highest MV2 discriminant, $p_{\text{T},b}^0 + p_{\text{T},b}^1$. The expected significance for a potential signal scenario with $(m_{\tilde{t}_1}, m_{\tilde{\chi}_1^\pm}, m_{\tilde{\chi}_1^0}) = (400, 100, 50)\,\text{GeV}$ reaches its maximum when applying a $p_{\text{T},b}^0 + p_{\text{T},b}^1 > 300\,\text{GeV}$ requirement. Figure 11.25b shows the simulated distribution of transverse momentum of the second jet leading in p_{T}, p_{T}^1. The SM contributions reach their maximum already for $p_{\text{T}}^1 < 150\,\text{GeV}$ where the signal contributions are negligible. Thus, a requirement of $p_{\text{T}}^1 > 150\,\text{GeV}$ is added to the SR optimised for the $(m_{\tilde{t}_1}, m_{\tilde{\chi}_1^\pm}, m_{\tilde{\chi}_1^0}) = (700, 100, 50)\,\text{GeV}$ scenario. Applying the basic requirements summarised in Table 11.1 as well as $p_{\text{T},b}^0 + p_{\text{T},b}^1 > 300\,\text{GeV}$, the simulated distributions of the transverse momentum of the fourth jet leading in p_{T}, p_{T}^3, $m_{\text{T}}^{b,\text{min}}$ and $m_{\text{T}}^{b,\text{max}}$ depicted in Fig. 11.26 are obtained. The maximum expected significance for the $(m_{\tilde{t}_1}, m_{\tilde{\chi}_1^\pm}, m_{\tilde{\chi}_1^0}) = (400, 100, 50)\,\text{GeV}$ signal scenario can be reached by requiring $p_{\text{T}}^3 > 100\,\text{GeV}$, $m_{\text{T}}^{b,\text{min}} > 250\,\text{GeV}$ and $m_{\text{T}}^{b,\text{max}} > 300\,\text{GeV}$.

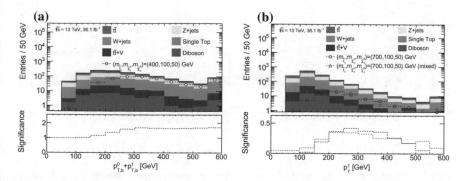

Fig. 11.25 Simulated distributions $p_{T,b}^0 + p_{T,b}^1$ (a) and p_T^1 (b) after requiring the presence of at least two b-jets and $m_T^{b,\min} > 200$ GeV in addition to the basic requirements summarised in Table 11.1. The bottom panel shows the expected significance for the signal scenarios with $(m_{\tilde{t}_1}, m_{\tilde{\chi}_1^\pm}, m_{\tilde{\chi}_1^0}) = (400, 100, 50)$ GeV and $(m_{\tilde{t}_1}, m_{\tilde{\chi}_1^\pm}, m_{\tilde{\chi}_1^0}) = (700, 100, 50)$ GeV depending on a potential cut value on the corresponding variable

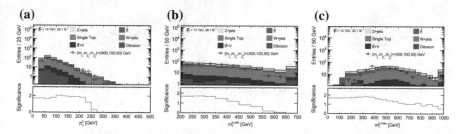

Fig. 11.26 Simulated distributions of p_T^3 (a), $m_T^{b,\min}$ (b) and $m_T^{b,\max}$ (c) after requiring the presence of at least two b-jets, $m_T^{b,\min} > 200$ GeV and $p_{T,b}^0 + p_{T,b}^1 > 300$ GeV in addition to the basic requirements summarised in Table 11.1. The bottom panel shows the expected significance for the signal scenario with $(m_{\tilde{t}_1}, m_{\tilde{\chi}_1^\pm}, m_{\tilde{\chi}_1^0}) = (400, 100, 50)$ GeV depending on a potential cut value on the corresponding variable

Applying all requirements obtained by the optimisation for the $(m_{\tilde{t}_1}, m_{\tilde{\chi}_1^\pm}, m_{\tilde{\chi}_1^0}) = (400, 100, 50)$ GeV signal scenario, the simulated distributions of the transverse momenta of the second (p_T^1) and fifth (p_T^4) jet leading in p_T as well as $\Delta R(b, b)$ are shown in Fig. 11.27. The expected significance depending on a requirement on p_T^1 is maximum for requiring $p_T^1 > 150$ GeV. For p_T^4, the maximum expected significance is reached for requiring $p_T^4 > 150$ GeV. However, the signal simulation used for the optimisation has poor statistics in this region. A requirement of $p_T^4 > 60$ GeV can be made without suffering from signal statistics. The expected significance depending on a requirement made on $\Delta R(b, b)$ rises at $\Delta R(b, b) > 0.8$ and reaches its maximum at $\Delta R(b, b) \sim 1.2$ (cf. Fig. 11.27(c)). In order to align the SR for the $(m_{\tilde{t}_1}, m_{\tilde{\chi}_1^\pm}, m_{\tilde{\chi}_1^0}) = (400, 100, 50)$ GeV signal scenario with the $(m_{\tilde{t}_1}, m_{\tilde{\chi}_1^\pm}, m_{\tilde{\chi}_1^0}) = (700, 100, 50)$ GeV, a requirement on $\Delta R(b, b) > 0.8$ is made.

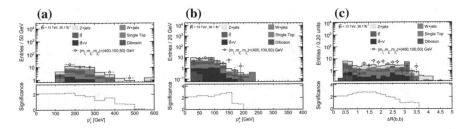

Fig. 11.27 Simulated distributions of p_T^1 (**a**), p_T^4 (**b**) and $\Delta R\,(b, b)$ (**c**) after applying all requirements discussed in Fig. 11.26. The bottom panel shows the expected significance for the signal scenario with $(m_{\tilde{t}_1}, m_{\tilde{\chi}_1^\pm}, m_{\tilde{\chi}_1^0}) = (400, 100, 50)\,\mathrm{GeV}$ depending on a potential cut value on the corresponding variable

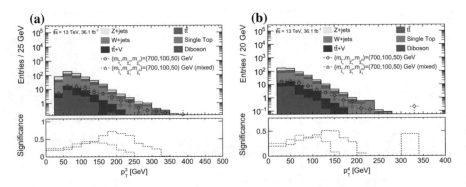

Fig. 11.28 Simulated distributions of p_T^3 (**a**) and p_T^4 (**b**) after requiring the presence of at least two b-jets, $m_\mathrm{T}^{b,\min} > 200\,\mathrm{GeV}$ and $p_\mathrm{T}^1 > 150\,\mathrm{GeV}$ in addition to the basic requirements summarised in Table 11.1. The bottom panel shows the expected significance for the signal scenario with $(m_{\tilde{t}_1}, m_{\tilde{\chi}_1^\pm}, m_{\tilde{\chi}_1^0}) = (700, 100, 50)\,\mathrm{GeV}$ depending on a potential cut value on the corresponding variable

For the signal scenario $(m_{\tilde{t}_1}, m_{\tilde{\chi}_1^\pm}, m_{\tilde{\chi}_1^0}) = (700, 100, 50)\,\mathrm{GeV}$, the simulated distributions of p_T^3 and p_T^4 which are shown in Fig. 11.28 suggest to require $p_\mathrm{T}^3 > 80\,\mathrm{GeV}$ and $p_\mathrm{T}^4 > 60\,\mathrm{GeV}$ in order to preserve sensitivity for the mixed decay scenario.

The remaining requirements for the SR targeting the $(m_{\tilde{t}_1}, m_{\tilde{\chi}_1^\pm}, m_{\tilde{\chi}_1^0}) = (700, 100, 50)\,\mathrm{GeV}$ signal scenarios can be seen in Fig. 11.29 where all requirements are applied so far and the simulated distributions of $p_{\mathrm{T},b}^0 + p_{\mathrm{T},b}^1$, $\Delta R\,(b, b)$, $m_\mathrm{T}^{b,\min}$ and $m_\mathrm{T}^{b,\max}$ are shown. All requirements made for SRD are summarised in Table 11.6.

Figure 11.30 shows the distribution of the distance between the first two b-jets with the highest MV2 discriminant, $\Delta R\,(b, b)$, after applying all cuts of SRD-low except $\Delta R\,(b, b)$ itself. The most dominant SM contributions in SRD-low (SRD-high) are arising from Z+jets and W+jets (single top quark) production, respectively. The expected yields in all SRs defined in Table 11.6 for a dataset of $36.1\,\mathrm{fb}^{-1}$ are shown in Table 11.7.

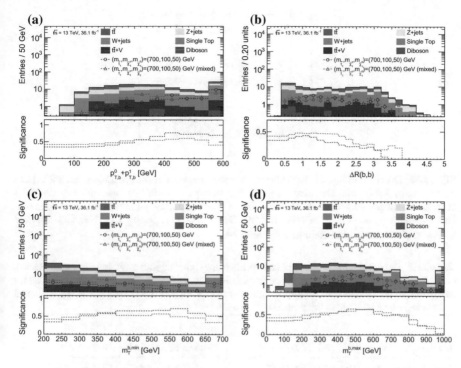

Fig. 11.29 Simulated distributions of $p_{\mathrm{T},b}^0 + p_{\mathrm{T},b}^1$ (**a**), $\Delta R\,(b,b)$ (**b**), $m_{\mathrm{T}}^{b,\min}$ (**c**) and $m_{\mathrm{T}}^{b,\max}$ (**d**) after applying all requirements discussed in Fig. 11.28. The bottom panel shows the expected significance for the signal scenario with $(m_{\tilde{t}_1}, m_{\tilde{\chi}_1^\pm}, m_{\tilde{\chi}_1^0}) = (700, 100, 50)\,\mathrm{GeV}$ depending on a potential cut value on the corresponding variable

Table 11.6 Signal region selection targeting models involving $\tilde{t}_1 \rightarrow b\tilde{\chi}_1^\pm$ decays in addition to the requirements presented in Table 11.1

Variable	SRD-low	SRD-high
$\left\|\Delta\phi\left(\mathrm{jet}^{0,1,2}, E_{\mathrm{T}}^{\mathrm{miss}}\right)\right\|$	>0.4	
$E_{\mathrm{T}}^{\mathrm{miss}}$	>250 GeV	
NJets	≥ 5	
b-tagged jets	≥ 2	
$\Delta R\,(b, b)$	>0.8	
τ-veto	Yes	
jet p_{T}^1	>150 GeV	
jet p_{T}^3	>100 GeV	>80 GeV
jet p_{T}^4	>60 GeV	
$m_{\mathrm{T}}^{b,\min}$	>250 GeV	>350 GeV
$m_{\mathrm{T}}^{b,\max}$	>300 GeV	>450 GeV
b-jet $p_{\mathrm{T},b}^0 + p_{\mathrm{T},b}^1$	>300 GeV	>400 GeV

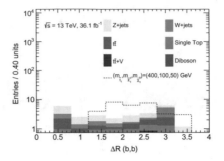

Fig. 11.30 Distribution of $\Delta R\,(b, b)$ after applying all requirements of SRD-low (cf. Table 11.6) but the one on $\Delta R\,(b, b)$ scaled to an integrated luminosity of $36.1\,\mathrm{fb}^{-1}$ without drawing statistical or systematic uncertainty bands. In addition to the SM processes, also simulated top squark pair production with $(m_{\tilde{t}_1}, m_{\tilde{\chi}_1^\pm}, m_{\tilde{\chi}_1^0}) = (400, 100, 50)\,\mathrm{GeV}$ and $(m_{\tilde{t}_1}, m_{\tilde{\chi}_1^\pm}, m_{\tilde{\chi}_1^0}) = (700, 100, 50)\,\mathrm{GeV}$ is shown. The rightmost bin includes overflow events

Table 11.7 Expected number of simulated events in the signal regions targeting top squark decays involving charginos scaled to an integrated luminosity of $36.1\,\mathrm{fb}^{-1}$. For the signal scenarios used for the optimisation of the SRs, the expected significance is given. For the SM processes, the statistical and the experimental systematic uncertainties are shown

	SRD-low	SRD-high
$(m_{\tilde{t}_1}, m_{\tilde{\chi}_1^\pm}, m_{\tilde{\chi}_1^0}) = (700, 200, 100)\,\mathrm{GeV}$	$13.66 \pm 1.32\,(1.3\sigma)$	$10.50 \pm 1.15\,(2.1\sigma)$
$(m_{\tilde{t}_1}, m_{\tilde{\chi}_1^\pm}, m_{\tilde{\chi}_1^0}) = (700, 200, 100)\,\mathrm{GeV}$ $(\mathcal{BR} = 50\%)$	$15.21 \pm 2.18\,(1.5\sigma)$	$6.85 \pm 1.03\,(1.4\sigma)$
$(m_{\tilde{t}_1}, m_{\tilde{\chi}_1^\pm}, m_{\tilde{\chi}_1^0}) = (700, 100, 50)\,\mathrm{GeV}$	$7.64 \pm 0.98\,(0.7\sigma)$	$6.14 \pm 0.94\,(1.3\sigma)$
$(m_{\tilde{t}_1}, m_{\tilde{\chi}_1^\pm}, m_{\tilde{\chi}_1^0}) = (700, 100, 50)\,\mathrm{GeV}$ $(\mathcal{BR} = 50\%)$	$10.08 \pm 1.17\,(1.0\sigma)$	$5.52 \pm 0.88\,(1.1\sigma)$
$(m_{\tilde{t}_1}, m_{\tilde{\chi}_1^\pm}, m_{\tilde{\chi}_1^0}) = (400, 100, 50)\,\mathrm{GeV}$	$32.75 \pm 6.05\,(2.9\sigma)$	$3.50 \pm 1.58\,(0.7\sigma)$
$t\bar{t}$	$3.43 \pm 0.37^{+0.609}_{-0.695}$	$1.04 \pm 0.20^{+0.230}_{-0.220}$
Z+jets	$6.68 \pm 0.44^{+1.126}_{-1.051}$	$3.10 \pm 0.27^{+0.580}_{-0.658}$
W+jets	$4.78 \pm 2.68^{+2.751}_{-2.881}$	$0.84 \pm 0.16^{+0.309}_{-0.348}$
Single Top	$3.30 \pm 0.47^{+0.773}_{-0.831}$	$1.30 \pm 0.22^{+0.183}_{-0.182}$
$t\bar{t} + V$	$3.06 \pm 0.31^{+0.432}_{-0.411}$	$1.06 \pm 0.15^{+0.222}_{-0.214}$
Diboson	$0.09 \pm 0.05^{+0.022}_{-0.166}$	$0.48 \pm 0.42^{+0.856}_{-0.728}$
Total SM	$21.35 \pm 2.80^{+3.785}_{-4.008}$	$7.83 \pm 0.63^{+1.634}_{-1.691}$

11.4   Background Estimation

In order to determine the normalisation of the dominating SM contributions in all SRs from comparisons between data and simulation, CRs are designed to be as close as possible to the event topologies in the SRs but still disjoint and with low signal

Fig. 11.31 Schematic Feynman graphs for SM top quark pair production (**a**) and direct $\tilde{t}_1$ pair production (**b**) with subsequent semi-leptonic decays. Note that both graphs result in the same final state

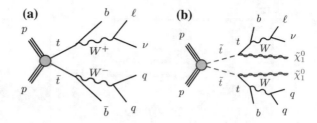

contamination. Solely, the normalisation of diboson production is entirely taken from simulation since it is one of the minor contributions throughout all SRs.

Top Quark Pair Production

Since the fully-hadronic $t\bar{t}$ decay is strongly suppressed by the $E_T^{miss} > 250\,\text{GeV}$ requirement, top quark pair production contributes to SRs when the charged lepton from the semi-leptonic $t\bar{t}$ decay (cf. Fig. 11.31a) is not reconstructed. This can be the case when the lepton is either out of the detector's acceptance, is mis-identified as a jet or is a hadronically decaying τ-lepton. The latter is the dominant process for all SRs. In order to define a CR for top quark pair production, one thus requires one isolated electron or muon with a transverse momentum of $p_T > 20\,\text{GeV}$ accounting for the semi-leptonic $t\bar{t}$ decay. For retaining a phase space as close as possible to the SRs, all basic requirements listed in Table 11.1 are kept except for the lepton veto which is replaced by the isolated lepton requirement and the requirements on $E_T^{miss,track}$ and $\left| \Delta\phi \left(E_T^{miss}, E_T^{miss,track} \right) \right|$ for suppressing multi-jet production, because the lepton requirement is already accounting for that. Due to the lepton being the substitution of a reconstructed jet from the SRs, it is treated as a non-b-tagged jet, in terms of being taken into account in the number of jets as well as in the requirements on their transverse momenta as summarised in Table 11.1. Since $m_T^{b,min}$ was introduced as the variable with the highest discrimination against $t\bar{t}$ production, the $m_T^{b,min} > 200\,\text{GeV}$ requirement is lowered to $m_T^{b,min} > 100\,\text{GeV}$ in order to enrich the purity of top quark pair production. However, requiring one isolated lepton and a similar selection as in the SRs targeting fully-hadronic top squark decays, results in a phase space potentially populated with decays of top squark pairs with one W boson decaying into lepton and neutrino (cf. Fig. 11.31b). In order to avoid a possible contamination of top squark pair production with subsequent semi-leptonic decays, the transverse mass between the lepton and the missing transverse energy is exploited a second time. Comparing pair production of SM top quarks and their superpartners (cf. Fig. 11.31), for top quark pair production the missing transverse energy arises from the neutrino only whereas for top squark pair production, also the neutralinos are contributing. As for a decay of a massive particle, the transverse mass between the decay products have a kinematic endpoint at the particle mass, the transverse mass between the lepton and the missing transverse energy, $m_T \left(\ell, E_T^{miss} \right)$, has its endpoint at the W boson mass for top quark pair production whereas at much higher values for simulated top squark pair production (cf. Fig. 11.32a). Thus, for all CRs containing exactly one isolated

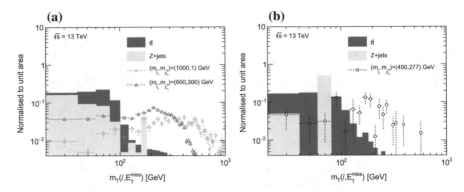

Fig. 11.32 Normalised distributions of $m_T\left(\ell, E_T^{\mathrm{miss}}\right)$ after requiring one charged lepton, all basic requirements listed in Table 11.1 except for $E_T^{\mathrm{miss,track}}$ and $\left|\Delta\phi\left(E_T^{\mathrm{miss}}, E_T^{\mathrm{miss,track}}\right)\right|$ for Z+jets and $t\bar{t}$ production. The lepton is treated as a non-b-tagged jet. In (**a**), where additionally $m_T^{b,\mathrm{min}} >$ 100 GeV is required, the signal scenarios used for the optimisation of SRA and SRB are shown. In (**b**), the signal scenario for the optimisation of SRC is shown

lepton, $m_T\left(\ell, E_T^{\mathrm{miss}}\right) < 100$ GeV is required. Furthermore, $m_T\left(\ell, E_T^{\mathrm{miss}}\right)$ helps to suppress several SM processes with particle decays involving neutrinos which are resulting in small values of $m_T\left(\ell, E_T^{\mathrm{miss}}\right)$ (cf. Figure 11.32a). Thus, additionally a lower requirement of $m_T\left(\ell, E_T^{\mathrm{miss}}\right) > 30$ GeV is made.

In order avoid unnecessary extrapolations from the $t\bar{t}$ CRs to the SRs, for every SR targeting high top squark masses (cf. Table 11.2) a distinct CR is defined by keeping the requirements on the reclustered jet masses as well as the requirements on $\Delta R\,(b, b)$ and $m_T^{b,\mathrm{max}}$ for SRB. For SRA, the requirements on the missing transverse energy are loosened by 200 GeV (150 GeV for SRA-TT) and the requirements on m_{T2} are dropped. Figure 11.33 shows the distributions of E_T^{miss} and $m_T^{b,\mathrm{min}}$ for both data and MC simulation with all requirements but the ones on E_T^{miss} and $m_T^{b,\mathrm{min}}$ applied. No difference in the shape between data and MC for both E_T^{miss} and $m_T^{b,\mathrm{min}}$ is observed justifying that an extrapolation in both variables from CR to SR can be made.

As explained in the following section about the single top quark production CR, an additional requirement on the minimal distance between the lepton and the two b-jets with the highest MV2 discriminant, $\Delta R(b_{0,1}, \ell)_{\mathrm{min}} < 1.5$, is made in order to be disjoint with respect to the single top quark production CR.

For the $t\bar{t}$ CR targeting SRC, the requirement on $m_T\left(\ell, E_T^{\mathrm{miss}}\right)$ is lowered to $m_T\left(\ell, E_T^{\mathrm{miss}}\right) < 80$ GeV to minimise the signal contamination (cf. Fig. 11.32b). In order to avoid that the $t\bar{t}$ normalisation factor targeting SRC is suffering from a high statistical uncertainty, the SR requirement on $\Delta\phi_{\mathrm{ISR},E_T^{\mathrm{miss}}}$ is completely dropped and the requirement on $p_T^{4,\mathrm{S}}$ is loosened to $p_T^{4,\mathrm{S}} > 40$ GeV. For the same reason, in the $t\bar{t}$ CR targeting the SRD selections, the requirements on $m_T^{b,\mathrm{max}}$ and p_T^3 were loosened to $m_T^{b,\mathrm{max}} > 100$ GeV and $p_T^3 > 80$ GeV, respectively, and the requirement on p_T^4 was

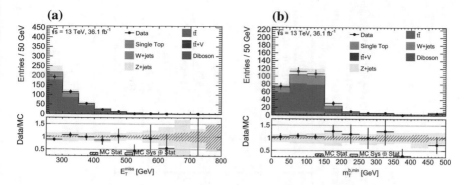

Fig. 11.33 Distributions of $E_{\mathrm{T}}^{\mathrm{miss}}$ and $m_{\mathrm{T}}^{b,\mathrm{min}}$ scaled to an integrated luminosity of 36.1 fb^{-1} after applying all requirements of the $t\bar{t}$ CR except for the ones on $E_{\mathrm{T}}^{\mathrm{miss}}$ and $m_{\mathrm{T}}^{b,\mathrm{min}}$. The yellow bands show the quadratic sum of the statistical and experimental systematic uncertainties

dropped completely. The final CR requirements for estimating the normalisation of $t\bar{t}$ production are summarised in Appendix B.5.

Table 11.8 summarises the event yields after applying all CR requirements on both data and simulated SM backgrounds as well as the purity of the $t\bar{t}$ contribution and the naive normalisation obtained by requiring

$$N_{\mathrm{Data}}^{\mathrm{CR}} = \mu_{t\bar{t}}^{\mathrm{CR}} \cdot N_{t\bar{t}}^{\mathrm{CR}} + \left(N_{\mathrm{MC}}^{\mathrm{CR}} - N_{t\bar{t}}^{\mathrm{CR}}\right) \quad . \tag{11.5}$$

The purity is above 84% throughout all $t\bar{t}$ CRs and the normalisation factor is in effect one considering the quadratic sum of the statistical and experimental systematic uncertainties for all regions apart from CRTC. Due to the requirement on $m_{\mathrm{T}}\left(\ell, E_{\mathrm{T}}^{\mathrm{miss}}\right)$, the contamination of potential top squark pair production is below 8% in all regions and for all simulated signal scenarios not excluded by previous ATLAS searches.

Production of Z Bosons in Association with Heavy-Flavour Jets

The production of a Z boson and additional jets contributes to the events in the SRs if the Z boson decays into a pair of neutrinos, causing missing transverse energy and if one or more heavy-flavour jets are present in the final state. To estimate the contribution of the $Z \rightarrow \nu\bar{\nu}$ decay, $Z \rightarrow \ell^+\ell^-$ events are used because they can be selected with a high purity compared to the other SM processes. However, due to the strict requirements on $E_{\mathrm{T}}^{\mathrm{miss}}$ in the SRs, the $Z(\rightarrow \nu\bar{\nu})$+jets contribution is primarily consisting of strongly boosted Z bosons for which the statistics of $Z \rightarrow \ell^+\ell^-$ decays is rather low. As for the $t\bar{t}$ CR, at least four jets two of which are b-tagged are required. Also the requirements on the jet transverse momenta used in the common preselection (cf. Table 11.1) are applied. Since two oppositely charged leptons are required, the electron and muon triggers with the lowest unprescaled p_{T}-threshold available are used as summarised in Table 11.9. In the case of $Z \rightarrow \mu^+\mu^-$ decays, the muon which fired the trigger is required to satisfy $|\eta| < 2.4$ in order to be within

Table 11.8 Composition of SM processes in control regions targeting $t\bar{t}$ production for 36.1 fb^{-1} of pp collision data. None of the simulated SM predictions is normalised using normalisation factors derived from a simultaneous fit (detailed in Sect. 11.6). The purities of the simulated $t\bar{t}$ contributions as well as the normalisation factors computed with Eq. (10.1) are shown for all regions. For simulated events, the statistical and the experimental systematic uncertainties are shown. The uncertainty on the purity and the normalisation factor includes the quadratic sum of the statistical and experimental systematic uncertainties

	CRTA-TT	CRTA-TW	CRTA-T0	CRTB-TT	CRTB-TW	CRTB-T0	CRTC	CRTD
$t\bar{t}$	$107.68 \pm 3.33^{+21.001}_{-24.811}$	$205.47 \pm 3.11^{+32.906}_{-36.031}$	$97.39 \pm 1.49^{+12.999}_{-13.142}$	$100.15 \pm 3.17^{+20.894}_{-24.871}$	$427.72 \pm 5.93^{+68.925}_{-73.812}$	$507.55 \pm 6.21^{+68.937}_{-64.325}$	$57.42 \pm 2.45^{+8.514}_{-13.226}$	$191.73 \pm 3.99^{+28.600}_{-26.144}$
Z+jets	$0.04 \pm 0.02^{+0.022}_{-0.019}$	$0.01 \pm 0.01 \pm 0.007$	$0.02 \pm 0.02^{+0.019}_{-0.026}$	$0.05 \pm 0.02^{+0.022}_{-0.020}$	$0.01 \pm 0.01^{+0.018}_{-0.038}$	$0.18 \pm 0.15^{+0.081}_{-0.096}$	$0.09 \pm 0.07^{+0.126}_{-0.087}$	$0.12 \pm 0.07^{+0.122}_{-0.103}$
W+jets	$4.48 \pm 0.57^{+1.255}_{-1.066}$	$2.19 \pm 0.58^{+0.599}_{-1.061}$	$3.72 \pm 1.01^{+1.306}_{-1.373}$	$5.45 \pm 0.93^{+1.242}_{-1.119}$	$4.12 \pm 1.05^{+0.964}_{-1.140}$	$6.12 \pm 1.10^{+1.263}_{-1.416}$	$2.21 \pm 0.70^{+0.422}_{-0.312}$	$6.26 \pm 0.72^{+1.310}_{-1.379}$
Single Top	$8.48 \pm 0.78^{+0.893}_{-1.206}$	$6.31 \pm 0.92^{+1.675}_{-1.910}$	$5.96 \pm 0.91^{+2.181}_{-1.559}$	$9.54 \pm 0.75^{+1.410}_{-2.012}$	$14.57 \pm 1.59^{+2.774}_{-2.869}$	$19.02 \pm 1.49^{+2.489}_{-3.125}$	$2.90 \pm 0.66^{+0.486}_{-0.778}$	$24.40 \pm 2.34^{+2.957}_{-3.858}$
$t\bar{t}+V$	$1.80 \pm 0.23^{+0.292}_{-0.296}$	$2.03 \pm 0.22^{+0.370}_{-0.431}$	$1.12 \pm 0.17^{+0.145}_{-0.147}$	$1.63 \pm 0.22^{+0.315}_{-0.322}$	$3.51 \pm 0.30^{+0.559}_{-0.624}$	$3.38 \pm 0.30^{+0.365}_{-0.411}$	$1.24 \pm 0.24^{+0.229}_{-0.250}$	$2.73 \pm 0.24^{+0.410}_{-0.379}$
Diboson	0.00 ± 0.00	$0.02 \pm 0.01^{+0.014}_{-0.149}$	$0.00 \pm 0.00 \pm 0.000$	$-0.39 \pm 0.39^{+0.352}_{-0.659}$	$0.17 \pm 0.15^{+0.218}_{-0.163}$	$0.87 \pm 0.83^{+1.066}_{-1.091}$	$0.24 \pm 0.21^{+0.489}_{-0.387}$	$0.75 \pm 0.53^{+0.418}_{-0.364}$
Total MC	$122.10 \pm 3.49^{+22.222}_{-26.222}$	$216.03 \pm 3.30^{+34.754}_{-38.864}$	$108.20 \pm 2.02^{+15.361}_{-13.794}$	$116.42 \pm 3.42^{+22.383}_{-27.032}$	$450.10 \pm 6.24^{+72.056}_{-78.297}$	$537.12 \pm 6.54^{+72.351}_{-67.876}$	$64.11 \pm 2.66^{+9.259}_{-13.929}$	$226.00 \pm 4.72^{+32.110}_{-30.705}$
Data	131.00 ± 11.45	213.00 ± 14.59	95.00 ± 9.75	127.00 ± 11.27	418.00 ± 20.45	494.00 ± 22.23	40.00 ± 6.32	210.00 ± 14.49
Purity of $t\bar{t}$ [%]	88.19 ± 19.30	95.11 ± 17.23	90.01 ± 12.96	86.02 ± 20.32	95.03 ± 16.63	94.49 ± 12.83	89.57 ± 20.18	84.84 ± 12.31
Normalisation of $t\bar{t}$	1.08 ± 0.28	0.99 ± 0.19	0.86 ± 0.16	1.11 ± 0.30	0.92 ± 0.17	0.92 ± 0.13	0.58 ± 0.19	0.92 ± 0.16

Table 11.9 List of single electron and muon triggers with the lowest unprescaled p_T-threshold available during 2015 and 2016 data taking

Year	Lepton	Level-1 p_T threshold (GeV)	HLT p_T threshold (GeV)
2015	$e^{\pm}$	20	26
	$e^{\pm}$	22	62
	$e^{\pm}$	22	122
	$\mu^{\pm}$	15	21
	$\mu^{\pm}$	20	52.5
2016	$e^{\pm}$	22	28
	$e^{\pm}$	22	62
	$e^{\pm}$	22	142
	$\mu^{\pm}$	20	27.3
	$\mu^{\pm}$	20	52.5

the coverage of the muon trigger chambers. The minimum p_T requirement for both electrons and muons is set to $p_T > 28\,\text{GeV}$.

A pair of charged leptons of the same flavour but opposite charge and an invariant mass in a window of 5 GeV around the Z boson mass is required. Figure 11.34a shows the distribution of E_T^{miss} after applying the $Z \to \ell^+\ell^-$ described so far. An additional requirement of $E_T^{\text{miss}} < 50\,\text{GeV}$ completely removes the contributions of single top quark and top quark pair production and retains the region of phase space where the MC simulation is modelling the data with reasonable accuracy. Finally, the transverse momenta of the charged leptons is vectorially removed from the calculation of E_T^{miss} in order to mimic the $Z(\to \nu\bar{\nu})$ decay present in the SRs. The resulting missing transverse energy with invisible leptons is named $E_T^{\text{miss}'}$. The discriminating variables related to the missing transverse energy, such as $m_T^{b,\text{min}}$ or $m_T^{b,\text{max}}$ can also be recalculated using $E_T^{\text{miss}'}$ and are named $m_T^{b,\text{min}'}$ and $m_T^{b,\text{max}'}$ accordingly. The distribution of $E_T^{\text{miss}'}$ after requiring $E_T^{\text{miss}} < 50\,\text{GeV}$ is shown in Fig. 11.34b. A cut on $E_T^{\text{miss}'} > 100\,\text{GeV}$ significantly removes a significant amount of diboson and $t\bar{t} + V$ production which increases the purity of the Z+jets contribution. Furthermore, for low values of $E_T^{\text{miss}'}$ other SM processes creating two oppositely charged leptons, such as low mass resonances are removed and a better agreement between data and MC simulation can be achieved. Due to the lack of statistics unlike for top quark pair production not every SR targeting high top squark masses can be reflected in one separate CR. Thus, two CRs are defined shared by the TT and TW category and the T0 category of SRs SRA and SRB, respectively. Since the Z+jets contribution in SRC is negligible, no CR is defined. For SRD-low and SRD-high, one shared CR is defined by requiring $m_T^{b,\text{min}'} > 200\,\text{GeV}$ and $m_T^{b,\text{max}'} > 200\,\text{GeV}$ on top of the selection described above, in order to retain decent statistics. Table 11.10 shows the event yields after applying all Z+jets CR requirements on both data and simulated SM backgrounds as well as the purity and the naive normalisation computed as for

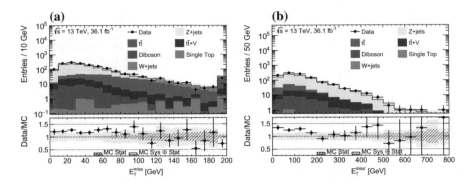

Fig. 11.34 Distributions of E_T^{miss} (**a**) and $E_T^{\text{miss}'}$ (**b**) scaled to an integrated luminosity of 36.1 fb^{-1} after requiring the presence of two oppositely charged leptons with an invariant mass not more than 5 GeV away from the Z boson mass. For (**b**), a cut on $E_T^{\text{miss}} < 50$ GeV is applied. The yellow bands show the quadratic sum of the statistical and experimental systematic uncertainties

Table 11.10 Composition of SM processes in control regions targeting Z+jets production for 36.1 fb^{-1} of pp collision data. None of the simulated SM predictions is normalised using normalisation factors derived from a simultaneous fit (detailed in Sect. 11.6). The purities of the simulated Z+jets contributions as well as the normalisation factors computed with Eq. (10.1) are shown for all regions. For simulated events, the statistical and the experimental systematic uncertainties are shown. The uncertainty on the purity and the normalisation factor includes the quadratic sum of the statistical and experimental systematic uncertainties

	CRZAB-TT-TW	CRZAB-TO	CRZD
$t\bar{t}$	$1.26 \pm 0.67^{+1.127}_{-1.006}$	$3.87 \pm 1.10^{+1.833}_{-1.628}$	$0.31 \pm 0.31^{+0.470}_{-0.477}$
Z+jets	$41.10 \pm 1.29^{+8.390}_{-9.066}$	$84.43 \pm 1.93^{+13.839}_{-13.474}$	$84.80 \pm 1.90^{+12.072}_{-13.367}$
W+jets	0.00 ± 0.00	0.00 ± 0.00	0.00 ± 0.00
Single Top	0.00 ± 0.00	0.00 ± 0.00	0.00 ± 0.00
$t\bar{t} + V$	$11.39 \pm 0.30^{+1.206}_{-1.329}$	$11.68 \pm 0.29^{+1.325}_{-1.339}$	$10.47 \pm 0.28^{+1.281}_{-1.214}$
Diboson	$3.05 \pm 0.65^{+1.089}_{-1.380}$	$4.13 \pm 0.68^{+0.633}_{-0.664}$	$3.80 \pm 0.52^{+1.357}_{-1.367}$
Total MC	$56.81 \pm 1.62^{+10.973}_{-11.883}$	$104.11 \pm 2.35^{+16.216}_{-15.237}$	$99.37 \pm 2.01^{+13.649}_{-15.288}$
Data	68.00 ± 8.25	119.00 ± 10.91	116.00 ± 10.77
Purity of Z+jets [%]	72.34 ± 15.44	81.10 ± 12.90	85.34 ± 13.38
Normalisation of Z+jets	1.27 ± 0.35	1.18 ± 0.24	1.20 ± 0.23

the $t\bar{t}$ CRs (cf. Table 11.8). All Z+jets CRs have a purity higher than 72% and a normalisation factor compatible with one considering the quadratic sum of the statistical and the experimental systematic uncertainties. Due to the lepton requirement and the upper cut on E_T^{miss}, the Z+jets CRs are not contaminated by any potential top squark signal.

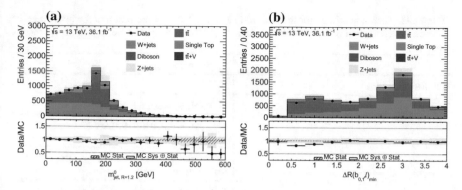

Fig. 11.35 Distributions of $m^0_{\mathrm{jet},R=1.2}$ (**a**) and $\Delta R(b_{0,1},\ell)_{\min}$ (**b**) scaled to an integrated luminosity of $36.1\,\mathrm{fb}^{-1}$ after requiring one charged lepton and all basic requirements listed in Table 11.1 except for $E_{\mathrm{T}}^{\mathrm{miss,track}}$ and $\left|\Delta\phi\left(E_{\mathrm{T}}^{\mathrm{miss}},E_{\mathrm{T}}^{\mathrm{miss,track}}\right)\right|$. The yellow bands show the quadratic sum of the statistical and experimental systematic uncertainties

Production of W Bosons in Association with Heavy-Flavour Jets

The production of a W boson and additional jets contributes in the SRs in case the W boson decaying into a charged lepton and a neutrino is boosted enough to give the neutrino the energy needed to fire the missing transverse energy trigger. In addition, one or more heavy-flavour jets have to be present in the final state to fulfil the b-jet requirements. As for SM $t\bar{t}$ production, W+jets can only contribute when the charged lepton is not reconstructed. Thus, the same basic requirements as for the $t\bar{t}$ CRs are applied. In contrary to $t\bar{t}$ production, where two bottom quarks are produced in the top quark decays, in W+jets events, the bottom quarks originate from gluon splitting in the underlying event. In order to suppress $t\bar{t}$ contributions, one single b-jet is required in the definition of the W+jets CR. Figure 11.35a shows the mass distribution of the $R=1.2$ reclustered jet leading in p_{T}, $m^0_{\mathrm{jet},R=1.2}$, after requiring all cuts listed in Table 11.11 but the ones on $m^0_{\mathrm{jet},R=1.2}$ and $\Delta R(b_{0,1},\ell)_{\min}$. Since no top quark is present in W+jets production, requiring $m^0_{\mathrm{jet},R=1.2} < 60\,\mathrm{GeV}$ further increases the purity of the W+jets contribution. The minimal distance between the lepton and the two b-jets with the highest MV2 discriminant, $\Delta R(b_{0,1},\ell)_{\min}$, is shown in Fig. 11.35b for the same selection. Requiring $\Delta R(b_{0,1},\ell)_{\min} > 2$ not only removes a significant amount of the $t\bar{t}$ contribution but also ensures that the W+jets CR is disjoint from the $t\bar{t}$ CR targeting SRC which has a ≥ 1 b-jet requirement.

Table 11.12 summarises the event yields after applying all W+jets CR requirements on both data and simulated SM backgrounds as well as the purity and the naive normalisation computed as for the $t\bar{t}$ CRs (cf. Table 11.8). The W+jets CR has a purity of 55% and a normalisation factor compatible with one considering the quadractic sum of the statistical and the experimental systematic uncertainties. The potential signal contamination is below 5%. Due to the lack of statistics one single CR has to be shared for all SRs.

Table 11.11 Selection criteria for the CR used to estimate the normalisation of the production of W bosons in association with heavy-flavour jets. $\mathscr{L}$ $[10^{34}\,\text{cm}^{-2}\text{s}^{-1}]$ is denoting the instantaneous luminosity

	Requirement
Number of leptons	$N_\ell = 1$
Trigger	E_T^{miss}, HLT threshold: 70 GeV (2015)
	E_T^{miss}, HLT threshold: 90 GeV (2016, $\mathscr{L} \le 1.02$)
	E_T^{miss}, HLT threshold: 100 GeV (2016, $1.02 < \mathscr{L} \le 1.13$)
	E_T^{miss}, HLT threshold: 110 GeV (2016, $\mathscr{L} > 1.13$)
Missing transverse energy	$E_T^{\text{miss}} > 250$ GeV
Number of jets	$N_{\text{jets}} \ge 4$
Jet transverse momenta (sorted by p_T)	>80 GeV, >80 GeV, >40 GeV, >40 GeV
Number of b-jets	$N_{b-\text{jets}} = 1$
$\left\lvert \Delta\phi\left(\text{jet}^{0,1}, E_T^{\text{miss}}\right)\right\rvert$	>0.4
Transverse mass between lepton and E_T^{miss}	30 GeV$< m_T\left(\ell, E_T^{\text{miss}}\right) < 100$ GeV
Mass of $R = 1.2$ reclustered jet	$m_{\text{jet}, R=1.2}^0 < 60$ GeV
Minimum distance between b-jets and lepton	$\Delta R(b_{0,1}, \ell)_{\text{min}} > 2$

Production of Single Top Quarks

The main part of the contribution from single top quark production in the SRs arises from the Wt-channel production. Comparing the final state objects of the Wt-channel production of single top quarks with top quark pair production (cf. Figs. 10.4 and 10.5, respectively), the only difference is the presence of a second bottom quark or W boson arising from the additional top quark decay. Accordingly, the production of single top quarks contributes to the SRs due to the same reasons as $t\bar{t}$ production and is also estimated in a CR using one isolated lepton. As one top quark is present, a $m_{\text{jet}, R=1.2}^0 > 120$ GeV requirement is applied. At least two b-jets are required to obtain a control region which is disjoint from the W+jets CR. The selection criterion $m_T^{b,\text{min}} > 200$ GeV suppresses $t\bar{t}$ contributions efficiently and ensures that the phase space of the single top CR is close to the one of the SRs. Figure 11.36 shows the distributions of $\Delta R(b_{0,1}, \ell)_{\text{min}}$ and $\Delta R(b, b)$, respectively, after the selection mentioned above is applied on top of the basic requirements from the $t\bar{t}$ CRs. The production of single top quarks tends to have larger $\Delta R(b_{0,1}, \ell)_{\text{min}}$ and $\Delta R(b, b)$ than $t\bar{t}$ production. The latter can be explained by the fact that the semi-leptonically decaying $t\bar{t}$ system is usually not produced at rest if it contributes with high E_T^{miss} to the SRs. Thus, the two b-jets from the top quark decays are rather located in the same hemisphere of the detector. For Wt-channel single top quark production, the

Table 11.12 Composition of SM processes in the control region targeting W+jets production for 36.1 fb^{-1} of pp collision data. None of the simulated SM predictions is normalised using normalisation factors derived from a simultaneous fit (detailed in Sect. 11.6). The purity of the simulated W+jets contribution as well as the normalisation factor computed with Eq. (10.1) are shown. For simulated events, the statistical and the experimental systematic uncertainties are shown. The uncertainty on the purity and the normalisation factor includes the quadratic sum of the statistical and experimental systematic uncertainties

	CRW
$t\bar{t}$	$241.41 \pm 4.38^{+19.374}_{-14.770}$
Z+jets	$3.03 \pm 0.44^{+0.822}_{-0.754}$
W+jets	$449.97 \pm 10.04^{+60.257}_{-54.400}$
Single Top	$107.13 \pm 3.61^{+10.940}_{-6.933}$
$t\bar{t} + V$	$1.68 \pm 0.21^{+0.280}_{-0.286}$
Diboson	$21.07 \pm 2.88^{+4.424}_{-4.075}$
Total MC	$824.28 \pm 11.90^{+78.317}_{-60.014}$
Data	878.00 ± 29.63
Purity of W+jets [%]	54.59 ± 5.39
Normalisation of W+jets	1.12 ± 0.17

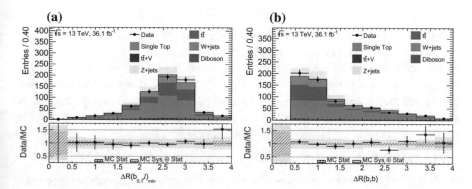

Fig. 11.36 Distributions of $\Delta R(b_{0,1}, \ell)_{\min}$ (**a**) and $\Delta R(b, b)$ (**b**) scaled to an integrated luminosity of 36.1 fb^{-1} after requiring one charged lepton, $m^0_{\text{jet},R=1.2} > 120$ GeV, $m^{b,\min}_{\text{T}} > 200$ GeV and all basic requirements listed in Table 11.1 except for $E^{\text{miss,track}}_{\text{T}}$ and $\left|\Delta\phi\left(E^{\text{miss}}_{\text{T}}, E^{\text{miss,track}}_{\text{T}}\right)\right|$. The yellow bands show the quadratic sum of the statistical and experimental systematic uncertainties

second b-jet usually arises from additional initial or final state radiation and is rather pointing back to back with respect to the b-jet from the top quark decay resulting in a larger $\Delta R(b, b)$. Therefore, the CR for single top quark production requires $\Delta R(b_{0,1}, \ell)_{\min} > 2$ and $\Delta R(b, b) > 1.5$.

Table 11.13 Composition of SM processes in the control region targeting single top quark production for 36.1 fb^{-1} of pp collision data. None of the simulated SM predictions is normalised using normalisation factors derived from a simultaneous fit (detailed in Sect. 11.6). The purity of the simulated single top quark contribution as well as the normalisation factor computed with Eq. (10.1) are shown. For simulated events, the statistical and the experimental systematic uncertainties are shown. The uncertainty on the purity and the normalisation factor includes the quadratic sum of the statistical and experimental systematic uncertainties

	CRST
$t\bar{t}$	$41.26 \pm 2.22^{+9.076}_{-8.517}$
Z+jets	$0.11 \pm 0.05^{+0.056}_{-0.072}$
W+jets	$24.77 \pm 2.22 \pm 6.352$
Single Top	$69.82 \pm 1.75^{+8.580}_{-8.774}$
$t\bar{t} + V$	$2.95 \pm 0.21^{+0.513}_{-0.463}$
Diboson	$2.93 \pm 1.26^{+0.862}_{-2.640}$
Total MC	$141.83 \pm 3.81^{+22.968}_{-23.734}$
Data	142.00 ± 11.92
Purity of single top [%]	49.23 ± 8.43
SF of single top	1.00 ± 0.23

Table 11.13 shows the event yields after applying all requirements of the single top quark CR on both data and simulated SM backgrounds as well as the purity and the naive normalisation computed as for the $t\bar{t}$ CRs (cf. Table 11.8). The single top quark CR has a purity of approximately 50% and a normalisation factor of one. The signal contamination if $\tilde{t}_1 \rightarrow t + \tilde{\chi}_1^0$ decays is below 8%, however, the decay scenario involving charginos shows a signal contamination of almost 14%. Due to the lack of statistics one single CR has to be shared for all SRs.

Top Quark Pair Production in Association with a Vector Boson

For SM $t\bar{t}$ production in association with a vector boson, the main contribution comes from $t\bar{t} + Z$ production (cf. Fig. 11.37a) with a subsequent decay of the Z boson into a pair of neutrinos. This type of background is the only irreducible background because it is a pure $t\bar{t} + E_T^{\text{miss}}$ final state as one would expect for a pair of top squarks decaying into top quarks and neutralinos. The estimation of $t\bar{t} + Z(\rightarrow \nu\bar{\nu})$ production using a $Z \rightarrow \ell^+\ell^-$ region as for Z+jets production is almost impossible since a selection optimised to select a Z boson is almost pure in contributions from Z+jets production and additional selection criteria targeting the $t\bar{t}$ pair lead to a large statistical uncertainty. Instead one makes use of the fact that processes with a Z boson are closely connected to processes with the production of a highly energetic photon γ^* via the electroweak symmetry. For sufficiently large momenta, where the kinematic constraints from the Z boson mass can be neglected, the ratio of the

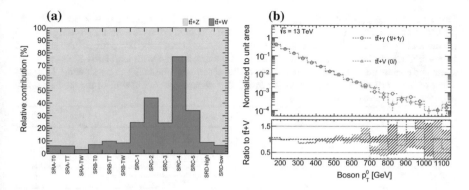

Fig. 11.37 Relative contributions of $t\bar{t} + Z$ and $t\bar{t} + W$ production in all SRs (**a**) and normalised distribution of transverse momentum of the Z boson (photon) for simulated $t\bar{t} + Z$ ($t\bar{t} + \gamma$) events at generator-level (**b**) requiring the presence of at least four jets and two b-jets as well as a lepton veto (exactly one lepton and one photon) as documented in Table B.4. Considering the statistical uncertainties, the shapes of the distributions for $t\bar{t} + Z$ and $t\bar{t} + \gamma$ production are in good agreement

production cross sections of $t\bar{t} + Z$ and $t\bar{t} + \gamma^*$ becomes proportional to the ratio of their coupling constants [37].

Figure 11.37b shows the transverse momentum of the boson for simulated $t\bar{t} + Z$ and $t\bar{t} + \gamma$ production, respectively, where the presence of at least four jets of which two are b-tagged is required. The shape of the distributions is in good agreement for transverse momenta above 150 GeV. Thus, a CR for $t\bar{t} + \gamma$ is defined to estimate the normalisation of $t\bar{t} + Z$ production in the SRs requiring exactly one charged lepton and one photon with $p_T > 150$ GeV. Furthermore, at least four jets of which two are b-tagged with the same p_T-requirements as for the basic preselection (cf. Table 11.1) are required. Since no requirement on the missing transverse energy is made, lepton triggers are used as summarised in Table 11.9. The lepton is required to have a transverse momentum higher than 28 GeV as for the CRs targeting Z+jets production.

Table 11.14 gives the event yields after applying all requirements of the $t\bar{t} + \gamma$ CR on both data and simulated SM backgrounds as well as the purity and the naive normalisation computed as for the $t\bar{t}$ CRs (cf. Table 11.8). The purity of $t\bar{t} + \gamma$ production in the CR is 80% and a normalisation factor of 1.2 is obtained. The $t\bar{t} + \gamma$ CR is not contaminated with any potential top squark signal and due to the lack of statistics, one single CR has to be shared for all SRs.

Estimation of Multi-jet Production

Although the contribution from multi-jet background in all SRs is strongly suppressed by the requirements on $\left|\Delta\phi\left(\text{jet}^{0,1}, E_T^{\text{miss}}\right)\right|$ or $\left|\Delta\phi\left(\text{jet}^{0,1,2}, E_T^{\text{miss}}\right)\right|$, respectively, as well as on $E_T^{\text{miss,track}}$ and $\left|\Delta\phi\left(E_T^{\text{miss}}, E_T^{\text{miss,track}}\right)\right|$ (cf. Table 11.1), its exact size has to be estimated in order to determine the final composition of SM processes for each SR. In addition to the impossibility to simulate multi-jet production with sufficient statistics in the SRs with strong requirements on E_T^{miss}, simulated multi-jet events

Table 11.14 Composition of SM processes in the control region targeting $t\bar{t} + V$ production for $36.1\,\text{fb}^{-1}$ of pp collision data. None of the simulated SM predictions is normalised using normalisation factors derived from a simultaneous fit (detailed in Sect. 11.6). The purity of the simulated $t\bar{t} + \gamma$ contribution as well as the normalisation factor computed with Eq. (10.1) are shown. For simulated events, the statistical and the experimental systematic uncertainties are shown. The uncertainty on the purity and the normalisation factor includes the quadratic sum of the statistical and experimental systematic uncertainties

	CRTTGamma
$t\bar{t}$	$16.23 \pm 2.29^{+2.266}_{-1.616}$
Z+jets	$0.67 \pm 0.17^{+0.127}_{-0.183}$
W+jets	$0.04 \pm 0.02^{+0.012}_{-0.014}$
Single Top	$2.11 \pm 0.80^{+0.436}_{-0.900}$
$t\bar{t} + V$	$2.34 \pm 0.25^{+0.340}_{-0.305}$
$t\bar{t} + \gamma$	$111.79 \pm 1.54^{+11.591}_{-11.315}$
$\gamma + V$	$5.90 \pm 0.63^{+1.181}_{-1.222}$
Diboson	0.00 ± 0.00
Total MC	$139.07 \pm 2.95^{+14.847}_{-14.345}$
Data	161.00 ± 12.69
Purity of $t\bar{t} + \gamma$ [%]	80.39 ± 8.82
Normalisation of $t\bar{t} + \gamma$	1.20 ± 0.17

are not expected to fully reproduce the non-Gaussian detector effects of measuring jets in data causing simulated multi-jet events to be poorly modelled. Therefore, the systematic uncertainties involved in modelling the calorimeter could be large [38]. In order to solve these problems, a method called *jet smearing* is applied. The method is based on the fact that multi-jet events contributing to the SRs arise from the mismeasurement of jet energies. Other possibilities that multi-jet production contributes in SRs would be that large $E_{\text{T}}^{\text{miss}}$ is reconstructed originating from additional pile-up jets, the removal of reconstruction ambiguities between leptons and jets or the the calculation of the $E_{\text{T}}^{\text{miss}}$ soft-term. However, for this topology, those sources are negligible.

A detailed overview of the original development of the method can be found in [39, 40]. As a starting point, the jet response $R_{\text{jet}} = p_{\text{T}}^{\text{reco}}/p_{\text{T}}^{\text{truth}}$ has to be measured in simulated multi-jet events by comparing the reconstructed jet energy $p_{\text{T}}^{\text{reco}}$ to the one at generator-level $p_{\text{T}}^{\text{truth}}$. Reconstructed jets which are spatially separated from other objects and matched within $\Delta R < 0.1$ to exactly one jet at generator-level are used. At generator-level, final state electrons, muons or neutrinos inside the jet cone are vectorially added to the four-momenta of the jet in order to account for genuine sources of $E_{\text{T}}^{\text{miss}}$ arising from heavy flavour decays. Analogously, reconstructed electrons and muons within the jet cone before removing reconstruction ambiguities are vectorially added to the four-momenta of the jet for obtaining a jet p_T as close as possible to the one at generator-level. The measured jet response is then tuned to collision data in two steps. First, the Gaussian core of the response is tuned by selecting back to back di-jet events low E_T^{miss} and fitting the asymmetry between the jet transverse momenta with a Gaussian distribution whose width gives an estimate of the jet energy resolution. Subsequently, the measured jet response can be tuned to match the widths observed in data [41]. The non-Gaussian tails of the jet response are tuned by selecting events with exactly three jets, where one of which is aligned collinear to the E_T^{miss}. Since this requirement ensures that the transverse momentum of one jet at generator-level, p_T^{truth}, is given by $p_T^{truth} = p_T^{reco} \pm E_T^{miss}$, a quantity sensitive to the tails of the jet response can be fitted in order to constrain the parameters describing the lower tail of the jet response [38].

Events with well-measured jets ($R_{jet} \sim 1$) are selected as so-called *seed events* for applying the jet smearing afterwards. Events containing well-measured jets could possibly be selected by requiring low missing transverse energies, however, it was found that a simple upper bound on E_T^{miss} for selecting seed events biases the average distribution of the smeared leading jet p_T towards lower values. High p_T jets are less likely to pass the seed selection and thus, rather low-p_T jets are used in the smearing which leads to a multi-jet prediction with a lower average leading jet p_T [39]. The E_T^{miss}-significance defined as the ratio between E_T^{miss} and the square root of the scalar sum of the transverse energies of all jets and soft objects in the event,

$$E_T^{miss,sign.} = \frac{E_T^{miss}}{\sqrt{\sum_{jets} E_T + \sum_{soft} E_T}} \quad , \tag{11.6}$$

is found to have a smaller dependence on the scalar sum of the transverse energies of all jets and soft objects. Thus, requiring an upper bound on $E_T^{miss,sign.}$ leads to a smaller bias. In order to further decrease the bias arising from the E_T^{miss} soft-term contributions, an additional parameter M is introduced for the definition of the modified E_T^{miss}-significance

$$E_{T,mod.}^{miss,sign.} = \frac{E_T^{miss} - M}{\sqrt{\sum_{jets} E_T + \sum_{soft} E_T}} \quad , \tag{11.7}$$

where $M = 8\,\text{GeV}$ was found to lead to the smallest bias [42]. For selecting seed events, the full 2015 dataset as well as the first $2.5\,\text{fb}^{-1}$ of 2016 were used requiring that at least on jet trigger has fired. Additionally, the presence of at least four jets, at least one b-jet and no lepton are required. Depending on the number of b-tagged jets N_b, the upper bound on the modified E_T^{miss}-significance is required to be

$$E_{T,mod.}^{miss,sign.} < 0.3 + 0.1 \cdot N_b \quad . \tag{11.8}$$

The value of 0.1 which is added to the cut value on $E_{\mathrm{T,mod.}}^{\mathrm{miss,sign.}}$ per b-jet is determined by measuring the shift of the average distribution of $E_{\mathrm{T,mod.}}^{\mathrm{miss,sign.}}$ in collision data events containing one b-jet with respect to events without any b-jet. The value of 0.3 is determined by optimising the agreement between collision data and the multi-jet prediction obtained in dedicated validation regions.

The transverse momenta of the jets of every seed event are smeared 5000 times using the tuned jet response and each time the missing transverse energy is reconstructed using the smeared transverse momenta. In cases where large fractions of the jet energy are not reconstructed by the calorimeter system, also the azimuthal angle of the jet can be mis-measured. Thus, the ϕ-coordinate needs to be smeared, too. The size of the variation in ϕ is estimated by fitting a Gaussian distribution centred around 0 to the $\Delta\phi$ distribution of di-jet events [41].

The jet smearing method is heavily relies on the assumptions that the tuned jet response measured in multi-jet simulation can be applied to jets of seed events without any problems and that there is no dependence of the jet response on other event properties such as jet multiplicity or the presence of heavy flavour jets. The multi-jet contribution is estimated in dedicated CRs where the cut on $\left|\Delta\phi\left(\mathrm{jet}^{0,1}, E_{\mathrm{T}}^{\mathrm{miss}}\right)\right|$ is inverted (cf. Appendix B.7). The method reproduces the distributions measured in data and leads to good agreement in all validation regions. Table 11.15 lists the estimated contributions of multi-jet production in all SRs.

Table 11.15 Expected yields of the multi-jet backgrounds in all SRs targeting top squark pair production estimated using the jet smearing technique

Region	Expected multi-jet yield
SRA-TT	0.21 ± 0.10
SRA-TW	0.14 ± 0.09
SRA-T0	0.12 ± 0.07
SRB-TT	1.54 ± 0.64
SRB-TW	1.01 ± 0.88
SRB-T0	1.79 ± 1.54
SRC-1	4.56 ± 2.38
SRC-2	1.58 ± 0.77
SRC-3	0.32 ± 0.17
SRC-4	0.04 ± 0.02
SRC-5	0.00 ± 0.00
SRD-low	1.12 ± 0.37
SRD-high	0.40 ± 0.15

11.5 Systematic Uncertainties

The simulation and subsequent reconstruction of all physics processes including both SM and potential supersymmetric particle production are associated with systematic uncertainties which are of experimental and theoretical origin. The precise estimation of the impact of these uncertainties on the event yields in all SRs and CRs as well as on the shape of the discriminating variables is a crucial ingredient for the statistical interpretation. The techniques employed in order to estimate the size of the systematic uncertainties are strongly dependent on the type of uncertainty, in especially concerning its experimental or theoretical origin.

Experimental Uncertainties

Uncertainties arising from the particle detection and event reconstruction mechanisms are referred to as experimental uncertainties and can be categorised into uncertainties which affect the event kinematics such as the energy calibration of jets or uncertainties of probabilistic origin such as the efficiency of reconstruction jet originating from a heavy flavour decay as a b-jet. Experimental uncertainties affecting the event kinematics are estimated by varying the corresponding calibration of interest by one standard deviation of its uncertainty and re-applying the selection to the newly calibrated objects. The difference between the yields obtained by applying the selection requirements with and without varying each kind of calibration separately, corresponds to the one standard deviation uncertainty used in the final statistical interpretation.

Experimental uncertainties originating from mismodelling of MC simulations or unforeseeable detector effects are usually estimated by varying the correction factors which are applied to all simulated events in order to correct for differences between recorded and simulated data by one standard deviation of their uncertainty (cf. the example of the reconstruction efficiency of muons presented in Chap. 8). Thus, as for uncertainties affecting the event kinematics, the variation of each correction factor applied results in a differing yield with respect to applying the nominal correction factors, which is propagated to the statistical interpretation accordingly.

Figure 11.38a shows the simulated distribution of the transverse momentum for the b-jet leading in p_T in the $t\bar{t}$ CR CRTB-T0 corresponding to 36.1 fb^{-1} of collision data. Figure 11.38b shows a breakdown of all experimental systematic uncertainties into their relative contributions. Thereby the experimental systematic uncertainties considered are:

- The energy calibration of jets is performed by comparing the calorimeter's response to the jet energy at generator level and its uncertainty depending on p_T and η is estimated using different in-situ techniques including the flavour composition of jets and pile-up effects [43]. The uncertainties are propagated to the statistical interpretation as a set of four (five) nuisance parameters (using the fast simulation framework). Figure 11.38b shows the quadratic sum of the uncertainties arising from the jet energy calibration labelled as *Jet energy scale* (JES). In addition to the JES, the resolution of the jet energy measurement is estimated by exploiting the transverse momentum balance in events containing jets with large transverse momenta [44]. The relative size of the systematic uncertainty of the *Jet energy resolution* (JER) measurement is also depicted in Fig. 11.38b.

- The uncertainty on the identification efficiency of jets containing b-hadrons which is also referred to as *b-tagging efficiency* is derived from MC generator modelling and normalisation uncertainties of the simulated events and experimental uncertainties related to detector effects and the reconstruction of the physics objects used for the efficiency measurement [45, 46].

- To account for the efficiency of the vertex reconstruction depending on the average number of interactions per proton bunch crossing, correction factors are applied to all simulated events to correct for differences between experimental and simulated data [47]. The uncertainties of the efficiency measurement which are mainly arising from the measurement of the average number of interactions per proton bunch crossing as well as discrepancies between data and MC simulation in the beam-spot size are propagated to the statistical interpretation as variations of the correction factors (cf. *Pile-up corrections* in Fig. 11.38b).

- In addition to the calibration uncertainties of the individual objects included in the calculation of the missing transverse energy, the uncertainties on the calculation of the E_T^{miss} soft-term are estimated by data-to-MC comparisons of the projections of $\mathbf{p}_T^{\mathrm{soft}}$ and $\mathbf{p}_T^{\mathrm{hard}}$ to $\mathbf{p}_T^{\mathrm{miss}}$ exploiting $Z \to \mu^+\mu^-$ decays with no additional jets in the events [48]. In Fig. 11.38b, the quadratic sum of all uncertainties related to the scale and resolution of the missing transverse energy is denoted as E_T^{miss}-*calibration*.

- To reject jets arising from pile-up interactions, requirements on the fraction of the total momentum of tracks in the jet which is associated with the primary vertex are made [49]. To retain a constant jet selection efficiency with respect to the pile-up interactions, a *Jet vertex tagging* (JVT) technique is exploited combining two pile-up-insensitive variables [50]. The JVT selection efficiency is estimated in $Z(\to \mu^+\mu^-)$+jets final states and correction factors are applied to all simulated events to correct for differences between experimental and simulated data. Uncertainties on the JVT efficiency measurement come from a potential mis-modelling of the $\Delta\phi(Z, \mathrm{jet})$ distribution and different efficiencies depending on the fragmentation models used in the MC simulations [50].

- In the CRs exploiting leptons, uncertainties on the determination of the energy scale of electrons arise from the fit of the invariant di-electron mass spectrum after selecting $Z \to e^+e^-$ decays, the choice of the gain used for the amplification of the electronic signals in the calorimeter cells and the modelling of the passive material traversed by the electrons before being stopped in the calorimeter [23]. For the measurement of the electron energy resolution using the Z boson resonance width, the uncertainties arise from the parametrisation of the energy resolution, contributions from pile-up jets and the modelling of the passive material. In Fig. 11.38b, the quadratic sum of the electron energy scale and resolution uncertainties is labelled as *Electron calibration*. In the case of CRs involving electrons, also the uncertainties of the correction factors applied to simulated events accounting for reconstruction, isolation and trigger efficiencies [20, 24] are propagated to the statistical interpretation.

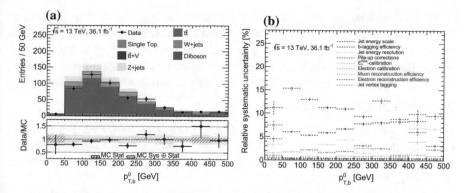

Fig. 11.38 Simulated distribution of transverse momentum for the b-jet leading in p_T in CRTB-T0 corresponding to $36.1\,\mathrm{fb}^{-1}$ of collision data **a** and corresponding relative contributions of experimental systematic uncertainties **b**. The yellow band overlaid on the SM prediction in the top panel as well as around the ratio between simulated top squark pair production and the SM prediction in the bottom panel, shows the quadratic sum of the statistical and all experimental systematic uncertainties. In the bottom panel, the fraction of is indicated as the black hashed area

- Uncertainties on the determination of the momentum scale and resolution of muons (cf. *Muon calibration* in Fig. 11.38b) exploiting $Z \rightarrow \mu^+\mu^-$ and $J/\psi \rightarrow \mu^+\mu^-$ decays originate from the choice of the width of the mass window used for the fit of the di-muon invariant mass spectrum, differences in the fitted parameters obtained from the Z or J/ψ resonance and the alignment of the muon chambers in the muon spectrometer [21]. As for electrons, in case of CRs involving muons, also the uncertainties of the correction factors applied to simulated events accounting for reconstruction, isolation and trigger efficiencies [21, 24] are considered in the statistical interpretation. Further details on the the determination of the muon reconstruction efficiencies can be found in Chap. 7.
- For the $t\bar{t} + \gamma$ CR used to estimate the normalisation of the $t\bar{t} + Z$ contribution in the SRs, uncertainties on the energy scale and resolution of photons have to be considered as well. In addition to the uncertainties described for the energy scale and resolution measurement of electrons, an uncertainty for photons converting to electron-positron pairs before reaching the calorimeter as well as wrong associations between ID tracks and calorimeter deposits induced by pile-up interactions or fake tracks is added [23]. Furthermore, the uncertainties of the correction factors applied to simulated events accounting for reconstruction and isolation efficiencies [51] are considered for the $t\bar{t} + \gamma$ CR.

Looking at Fig. 11.38b, the dominating experimental uncertainties arise from the determination of the jet energy scale and resolution as well as the uncertainties on the correction factors accounting for the b-tagging efficiency followed by the uncertainty on the pile-up correction factors. Looking at the size of the relative experimental uncertainties for all SRs (cf. Appendix B.8), a similar ordering can be found. This is expected because in the analysis one searches for final states with several jets including b-tagged ones and high missing transverse energy.

To estimate the systematic uncertainty of the jet smearing method, the upper bound for the modified $E_\mathrm{T}^\mathrm{miss}$-significance used for the seed selection (cf. Eq. (11.8)) is varied upwards to $E_\mathrm{T,mod.}^\mathrm{miss,sign.} < 0.6 + 0.2 \cdot N_b$ and downwards to $E_\mathrm{T,mod.}^\mathrm{miss,sign.} < 0.2 + 0.05 \cdot N_b$ and the maximum variation of the multi-jet estimate with respect to the nominal one is added in quadrature with a 30% uncertainty due to low tail uncertainties [1].

Theoretical Uncertainties

Since the optimisation of all SRs as well as the shape of the distributions of the discriminating variables are purely based on MC simulations, the uncertainties on the theoretical assumptions made when generating MC events must be estimated as precisely as possible.

For the SM processes where no dedicated CRs are used to estimate the normalisation of the process in the SRs as well as for potential supersymmetric processes, the total normalisation in the SRs is taken from simulation. Therefore the uncertainty in the production cross section of the process has to be added as an uncertainty on the normalisation.

Theoretical uncertainties affecting the shape of the distribution of a discriminating variable can arise from the choice of the PDF set, the factorisation and renormalisation scales, the model used for the parton showering or the extent of the emission of additional partons in the initial or final states.

Depending on the MC generator, the theoretical uncertainties are propagated by either event weights accounting for a variation of a parameter by its uncertainty or by generating new MC events where the theoretical parameter of interest is varied by its uncertainty. For deriving the theoretical uncertainties for a specific SM process x whose normalisation is estimated in a dedicated CR, the ratio of the predicted yields between the SR and the corresponding CR(x), which is also referred to as *transfer factor*,

$$T(x, v) = \frac{N_\mathrm{SR}}{N_\mathrm{CR}(x)} \quad,$$ (11.9)

is calculated for all possible variations v of the MC generator settings and the uncertainty on the transfer factor is propagated to the statistical interpretation. Theoretical uncertainties are derived individually for the SM processes with significant contributions to the SRs.

For SM $t\bar{t}$ production, the uncertainty on the choice of the MC generator $\sigma(t\bar{t},\mathrm{MC}$ gen.) is estimated by deriving the difference of the transfer factor between the using the nominal $t\bar{t}$ simulation generated using POWHEG-BOX 2 ($T(t\bar{t}$, nom.) and one generated in SHERPA 2.2.1 (cf. Table B.14 in Appendix),

$$\sigma(t\bar{t}, \mathrm{MCgen.}) = \frac{T(t\bar{t}, \mathrm{nom.}) - T(t\bar{t}, \mathrm{SHERPA\ 2.2.1})}{T(t\bar{t}, \mathrm{nom.})} \quad,$$ (11.10)

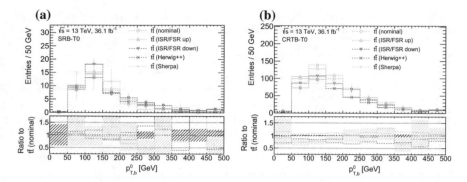

Fig. 11.39 Distribution of transverse momentum for the b-jet leading in p_T in SRB-T0 (**a**) and CRTB-T0 (**b**) for several simulations of $t\bar{t}$ production scaled to an integrated luminosity of 36.1 fb^{-1}

for all SRs and corresponding $t\bar{t}$ CRs. Compared to POWHEG-BOX 2, the modelling of additional partons in the final state is described more precisely in SHERPA 2.2.1 by exploiting a dedicated parton-shower model [52]. The first column of Table 11.16 shows the relative theoretical uncertainty on the choice of the MC generator obtained on the transfer factor for all SRs in which the difference of transfer factors is statistically meaningful.

For deriving the uncertainties on the choice of the model used for the parton showering of $t\bar{t}$ production, the difference of the transfer factor between using the nominal $t\bar{t}$ simulation where the parton showering is performed using PYTHIA 6.428 and a dedicated simulation showering using Herwig++ [53] (cf. Table B.14 in Appendix) is also calculated since PYTHIA 6.428 and Herwig++ are exploiting different parton shower-matrix element matching procedures [54]. The results obtained using Eq. (11.10) shown in the second column of Table 11.16.

The uncertainties on the extent of the radiation of additional partons in the initial or final states (ISR/FSR) are estimated by two dedicated $t\bar{t}$ simulations where the amount of radiation emitted in the initial or final state was set to half or twice the value of the nominal sample, respectively. The uncertainties on the transfer factors are then derived by

$$\sigma(t\bar{t}, \text{ISR/FSR}) = \frac{T(t\bar{t}, \text{ISR/FSRx2}) - T(t\bar{t}, \text{ISR/FSRx0.5})}{T(t\bar{t}, \text{ISR/FSRx2}) + T(t\bar{t}, \text{ISR/FSRx0.5})} \quad . \tag{11.11}$$

Figure 11.39 illustrates the effect of the different theoretical uncertainties for $t\bar{t}$ production on the distribution of transverse momentum of the b-jet leading in p_T in SRB-T0. The predicted event yields for each of the $t\bar{t}$ simulations correspond to the integral of the distributions shown in Fig. 11.39a and 11.39b for SRB-T0 and CRTB-T0, respectively.

Since the main part of the contribution from single top quark production in the SRs comes from the Wt-channel production, the theoretical uncertainties on single top

Table 11.16 Summary of relative theoretical uncertainties on $t\bar{t}$ production obtained on the transfer factor for all SRs. Only the uncertainties which are statistically meaningful are shown

Signal region	MC generator	Parton showering	ISR/FSR
SRA-TT	–	0.90	0.53
SRA-TW	1.0897	0.20	0.11
SRA-T0	0.3530	0.30	0.32
SRB-TT	0.3704	0.39	0.33
SRB-TW	0.1839	0.20	0.33
SRB-T0	0.4281	0.21	0.22
SRC-1	0.3780	0.05	0.20
SRC-2	0.0864	0.06	0.10
SRC-3	0.1639	0.11	0.04
SRC-4	0.1379	0.11	0.10
SRC-5	–	0.20	0.05
SRD-low	0.9204	0.62	0.28
SRD-high	0.8509	1.60	0.05

quark production are only estimated in the Wt-channel. Similarly to $t\bar{t}$ production, the theoretical uncertainties on the parton showering are derived by a Eq. (11.10) using a dedicated simulation where the parton showering is performed using Herwig++ instead of PYTHIA 6.428 (cf. Table B.14). Also, for the uncertainties on the extent of the radiation of additional partons in the initial or final states, the same approach as for $t\bar{t}$ production is used relying on two additional simulations where the amount of radiation emitted is varied (cf. Table B.14). To account for interference terms between SM $t\bar{t}$ and single top quark production, events are generated including all possible final states containing two W bosons and two bottom quarks. Comparing the predicted SR yields with the sum of $t\bar{t}$ and single top quark Wt production with an additional bottom quark, differences below 30% were observed for all SRs with meaningful statistics. Thus, a 30% uncertainty is applied to account for interference effects for the single top quark contributions throughout all SRs. For the statistical interpretation, the uncertainties arising from the parton showering model, the ISR/FSR variations and the interference effects are added quadratically and propagated as one systematic uncertainty as shown in Table 11.17.

For the production of Z and W bosons and additional jets, theoretical uncertainties on the choice of the renormalisation, factorisation and resummation scales are derived by varying those by a factor of two up- and downwards at generator-level exploiting LHE3 weights [55]. The differences in the predicted SR yields estimated by Eq. (11.11) are summed up quadratically and propagated into the statistical interpretation as one theoretical uncertainty for Z+jets and W+jets production, respectively (cf. Table 11.18). The uncertainty on the choice of the PDF set was found to be negligible.

Table 11.17 Quadratic sum of relative theoretical uncertainties on single top quark production obtained from the transfer factor for all SRs

Signal region	Relative theoretical uncertainty
SRA-TT	0.172
SRA-TW	0.157
SRA-T0	0.116
SRB-TT	0.100
SRB-TW	0.100
SRB-T0	0.108
SRC-1	0.116
SRC-2	0.142
SRC-3	0.121
SRC-4	0.091
SRC-5	0.282
SRD-low	0.112
SRD-high	0.103

Table 11.18 Quadratic sum of relative theoretical uncertainties on single Z+jets and W+jets production obtained on the transfer factor for all SRs. Only the uncertainties which are statistically meaningful are shown

Signal region	Z+jets	W+jets
SRA-TT	0.0448	0.0951
SRA-TW	0.0549	0.0803
SRA-T0	0.0285	0.0607
SRB-TT	0.0406	0.0912
SRB-TW	0.0407	0.0791
SRB-T0	0.0307	0.0325
SRC-1	–	0.1141
SRC-2	–	0.1254
SRC-3	–	0.1185
SRC-4	–	0.1073
SRC-5	–	0.0949
SRD-high	0.0219	0.0820
SRD-low	0.0247	0.0882

The theoretical uncertainties on $t\bar{t}$ production in association with a vector boson are estimated in the same way as those for Z+jets and W+jets production. The uncertainty on the choice of the PDF set is derived by comparing the PDFs available in NNPDF3.0NNLO. The quadratic sum of the relative theoretical uncertainties of $t\bar{t} + V$ production is found to be below 5% for all SRs. Thus, a relative uncertainty of 5% is applied on the $t\bar{t} + V$ contributions in all SRs.

For diboson production whose contribution in all SRs is negligible, a theoretical uncertainty of 50% of the predicted yields is propagated to the statistical interpretation.

Similar to the SM processes, the theoretical uncertainties on the signal simulations can be estimated by varying the modelling parameters including the production cross section by their uncertainties. The uncertainties obtained are added in quadrature and propagated as a single uncertainty for each SR and simulated dataset. The relative uncertainties are listed in Table B.18 in the Appendix.

11.6   Statistical Interpretation

To interpret the collision data recorded based on the MC predictions, the profile likelihood ratio test [56] is used. A short overview of the method is given below. More detailed information about the software packages used can be found in [57–60].

The number of expected events ν after applying the SR requirements can be expressed as $\nu = \mu \cdot s + b$ with the *signal strength* parameter μ and with the numbers of expected signal and background events, s and b, respectively. The number N_i of selected events in each bin i of the distribution of the final discriminating variable used for signal versus background separation follows a Poisson distribution

$$\text{Pois}(N_i | \mu \cdot s_i(\boldsymbol{\theta}) + b_i(\boldsymbol{\theta})) = \frac{(\mu \cdot s_i(\boldsymbol{\theta}) + b_i(\boldsymbol{\theta}))^{N_i}}{N_i!} \exp(-\mu \cdot s_i(\boldsymbol{\theta}) - b_i(\boldsymbol{\theta}))$$

(11.12)

where $\boldsymbol{\theta} = \left(\theta_j^{\text{sys}}, \theta^{\text{stat}} \right)$ represents a set of nuisance parameters taking into account the systematic uncertainties of the measurement via Gaussian probability density functions $\mathcal{G}(\theta^{\text{sys}} | \text{mean} = 0, \sigma = 1)$ and the statistical uncertainty θ^{stat} of each MC simulation.

The *likelihood function* $\mathcal{L}$ is given by

$$\mathcal{L}(N_{\text{obs}} | \mu, \boldsymbol{\theta}) = \prod_{i \in \text{bins}} \text{Pois}(N_i | \mu \cdot s_i(\boldsymbol{\theta}) + b_i(\boldsymbol{\theta})) \cdot \Gamma(\theta_i^{\text{stat}} | \beta_i) \cdot \prod_{\substack{j \in \text{systematic} \\ \text{uncertainties}}} \mathcal{G}(\theta_{ij}^{\text{sys}} | \text{mean} = 0, \sigma = 1)$$

(11.13)

where N_{obs} is the total number of data events satisfying the SR selection. Thus, the $\theta_j = \pm 1$ variations of the nuisance parameters correspond to $\pm 1\sigma$ variations of the parameters affected by the systematic uncertainties. The statistical uncertainties β_i for each bin i of a MC sample are modelled using an extended gamma function $\Gamma(\theta^{\text{stat}} | \beta_i)$ [61]. The number of expected background events b_i for each bin i is the sum of expected background events of all backgrounds which are taken into account,

$$b_i(\boldsymbol{\theta}) = \sum_{k \in \text{backgrounds}} \mu_k \cdot b_{i,k}(\boldsymbol{\theta}) , \qquad (11.14)$$

where μ_k are the transfer factors derived in the CRs.

By maximising the likelihood function depending on the parameters μ and $\boldsymbol{\theta}$, the global maximum likelihood estimators $\hat{\mu}$ and $\hat{\boldsymbol{\theta}}$ are determined.

For testing the compatibility of data and MC predictions, the profiled likelihood ratio [62]

$$\tilde{p}_\mu = -2 \cdot \ln \left(\frac{\mathcal{L}(N_{\text{obs}} | \mu, \hat{\boldsymbol{\theta}}_\mu)}{\mathcal{L}(N_{\text{obs}} | \hat{\mu}, \hat{\boldsymbol{\theta}})} \right) \qquad (11.15)$$

is used where $\hat{\boldsymbol{\theta}}_\mu$ is the maximum likelihood estimator of $\boldsymbol{\theta}$ for a fixed $\mu \in [0, \hat{\mu}]$. Under the background-only hypothesis H_0, the probability to observe data less compatible with H_0 than the actual measurement is given by the background-only p value

$$p_b = P(\tilde{p}_\mu > \tilde{p}_\mu^{\text{observed}} | H_0) \qquad (11.16)$$

where $P(\tilde{p}_\mu > \tilde{p}_\mu^{\text{observed}} | H_0)$ is the probability to calculate a profiled likelihood ratio greater than the one observed when assuming the background-only hypothesis H_0. On the other hand, under the signal-plus-background hypothesis H_μ, the probability to observe data less compatible with H_μ than the the actual measurement is given by the signal-plus-background p-value

$$p_{s+b} = P(\tilde{p}_\mu > \tilde{p}_\mu^{\text{observed}} | H_\mu) . \qquad (11.17)$$

In a first step, only the CRs are used to constrain the fit parameters and potential signal contaminations are neglected. Since the CRs estimating the normalisation of W+jets, single top quark and $t\bar{t} + \gamma$ production are used for the extrapolation to all SRs due to their limited statistics, it has to be decided which of the $t\bar{t}$ and Z+jets CR is used in the simultaneous fit to estimate the normalisation of $t\bar{t}$ and Z+jets production in the respective CRs. Since $m_{\text{jet}, R=1.2}^0 > 120\,\text{GeV}$ is required for the single top CR, the $t\bar{t}$ and Z+jets CRs targeting SRD are used because no requirement on $m_{\text{jet}, R=1.2}^0$ is made there.

The CR estimating the normalisation of single top quark production is mainly containing events with $m_{\text{jet}, R=1.2}^1 < 60\,\text{GeV}$ (cf. Fig. 11.40) and the E_T^{miss} requirement of the single top CR is closer to the one of the $t\bar{t}$ CRs targeting SRB, CRTB-T0 and CRZAB-T0 are used to normalise the $t\bar{t}$ and Z+jets contributions in the single top CR, respectively.

For the $t\bar{t} + \gamma$ CR estimating the normalisation of $t\bar{t} + Z$ production in the SRs, the same argument as for the single top quark production CR is applied.

The extrapolation of the normalisation factors obtained in the fit of the CRs are cross-checked by defining validation regions (VRs) which are kinematically similar

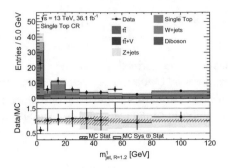

Fig. 11.40 Distribution of $m^1_{\text{jet},R=1.2}$ in the CR for single top quark production. The majority of the single top quark contribution satisfies $m^1_{\text{jet},R=1.2} < 60\,\text{GeV}$ which corresponds to the $R=1.2$ reclustered jet category T0. The yellow band shows the quadratic sum of the statistical and the experimental systematic uncertainties

Table 11.19 Fit results in SRA and SRB for an integrated luminosity of 36.1 fb^{-1}. The background normalisation parameters are obtained from the background-only fit in the CRs and are applied to the SRs. The uncertainties in the yields include statistical uncertainties and all systematic uncertainties defined in Sect. 11.5

	SRA-TT	SRA-TW	SRA-T0	SRB-TT	SRB-TW	SRB-T0
Observed	11	9	18	38	53	206
Total SM	8.6 ± 2.1	9.3 ± 2.2	18.7 ± 2.7	39.3 ± 7.6	52.4 ± 7.4	179 ± 26
$t\bar{t}$	$0.71^{+0.91}_{-0.71}$	$0.51^{+0.55}_{-0.51}$	1.31 ± 0.64	7.3 ± 4.3	12.4 ± 5.9	43 ± 22
Z+jets	2.5 ± 1.3	4.9 ± 1.9	9.8 ± 1.6	9.0 ± 2.8	16.8 ± 4.1	60.7 ± 9.6
W+jets	0.82 ± 0.15	0.8 ± 0.56	2.00 ± 0.83	7.8 ± 2.8	4.8 ± 1.2	25.8 ± 8.8
Single Top	1.20 ± 0.81	0.70 ± 0.42	2.9 ± 1.5	4.2 ± 2.2	5.9 ± 2.8	26 ± 13
$t\bar{t} + V$	3.16 ± 0.66	1.84 ± 0.39	2.60 ± 0.53	9.3 ± 1.7	10.8 ± 1.6	20.5 ± 3.2
Diboson	–	0.35 ± 0.26	–	0.13 ± 0.07	0.60 ± 0.43	1.04 ± 0.73
Multi-jet	0.21 ± 0.10	0.14 ± 0.09	0.12 ± 0.07	1.54 ± 0.64	1.01 ± 0.88	1.8 ± 1.5

to the CRs but closer to the SRs (cf. Appendix B.6). Tables 11.19, 11.20 and 11.21 show the observed event yields as well as the fit results for the SM contributions determined from a simultaneous fit to all CRs and SRs. No significant excess above the SM prediction is observed in any of the SRs (cf. Appendix B.9).

Figure 11.41 shows the distributions of $m_T^{b,\text{max}}$ for SRB-TW (a) and R_{ISR} for SRC1-5 (b) after the likelihood fit is performed. Since the data are in good agreement with the SM predictions and no excess above SM expectations can be observed, an exclusion fit is performed using both CRs and SRs to constrain the fit parameters. The signal contributions predicted by the simulations of top squark pair production for different top squark, neutralino (and chargino) masses (in case of $\tilde{t}_1 \rightarrow b + \tilde{\chi}_1^\pm$) are taken into account in both CRs and SRs by letting the signal strength parameter μ free but positive.

To protect exclusion limits against downward fluctuations of the data, the CL$_s$ method [63–65] is used based on the ratio CL$_s = p_{s+b}/p_b$. A 95% confidence level

Table 11.20 Fit results in SRC for an integrated luminosity of 36.1 fb^{-1}. The background normalisation parameters are obtained from the background-only fit in the CRs and are applied to the SRs. The uncertainties in the yields include statistical uncertainties and all systematic uncertainties defined in Sect. 11.5

	SRC-1	SRC-2	SRC-3	SRC-4	SRC-5
Observed	20	22	22	1	0
Total MC	20.6 ± 6.5	27.6 ± 4.9	18.9 ± 3.4	7.7 ± 1.2	0.91 ± 0.73
$t\bar{t}$	12.9 ± 5.9	22.1 ± 4.3	14.6 ± 3.2	4.91 ± 0.97	$0.63^{+0.70}_{-0.63}$
Z+jets	–	–	–	–	–
W+jets	0.80 ± 0.37	1.93 ± 0.49	1.91 ± 0.62	1.93 ± 0.46	0.21 ± 0.12
Single Top	1.7 ± 1.3	$1.2^{+1.4}_{-1.2}$	1.22 ± 0.69	0.72 ± 0.37	–
$t\bar{t} + V$	0.29 ± 0.16	0.59 ± 0.38	0.56 ± 0.31	0.08 ± 0.08	0.06 ± 0.02
Diboson	0.39 ± 0.33	$0.21^{+0.23}_{-0.21}$	0.28 ± 0.18	–	–
Multi-jet	4.6 ± 2.4	1.58 ± 0.77	0.32 ± 0.17	0.04 ± 0.02	–

Table 11.21 Fit results in SRD for an integrated luminosity of 36.1 fb^{-1}. The background normalisation parameters are obtained from the background-only fit in the CRs and are applied to the SRs. The uncertainties in the yields include statistical uncertainties and all systematic uncertainties defined in Sect. 11.5

	SRD-low	SRD-high
Observed	27	11
Total MC	25.1 ± 6.2	8.5 ± 1.5
$t\bar{t}$	3.3 ± 3.3	0.98 ± 0.88
Z+jets	6.9 ± 1.5	3.21 ± 0.62
W+jets	6.1 ± 2.9	1.06 ± 0.34
Single Top	3.8 ± 2.1	1.51 ± 0.74
$t\bar{t} + V$	3.94 ± 0.85	1.37 ± 0.32
Diboson	–	–
Multi-jet	1.12 ± 0.37	0.40 ± 0.15

(CL) upper limit on μ corresponds to a CL$_s$ value of $p_{s+b}/p_b \leq 0.05$. The upper limit on μ can be translated into an upper limit on the signal cross section σ_s using the relation $\nu = \mu \cdot s + b = L \cdot (\mu \cdot \sigma_s + \sigma_b)$ with the integrated luminosity L of the measurement.

Disjoint SRs, such as SRA-TT, SRA-TW and SRA-T0 are statistically combined by multiplying their likelihood functions. For overlapping SRs, the SR with the smallest CL$_s$ value is taken. After the regions are potentially combined, the region with the smallest CL$_s$ value is chosen to create an exclusion contour at 95% CL as a function of the top squark and neutralino masses assuming top squarks decaying with $\mathcal{BR}(\tilde{t}_1 \rightarrow t + \tilde{\chi}^0_1) = 100\%$ (cf. Fig. 11.42). The combination contains an additional SR which originally targets direct pair production of gluinos which subsequently decay into top quarks and their supersymmetric partners (cf. Appendix B.4). Top

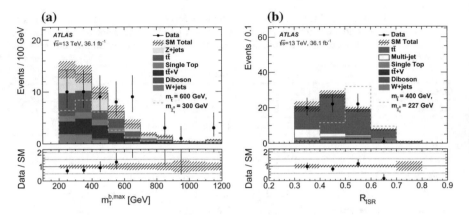

Fig. 11.41 Distributions of $m_T^{b,\max}$ for SRB-TW (**a**) and R_{ISR} for SRC1-5 (**b**) after the likelihood fit is performed. The stacked histograms show the SM prediction and the hatched uncertainty band around the SM prediction shows the MC statistical and detector-related systematic uncertainties. For each variable, the distribution for a representative signal point is shown [22]

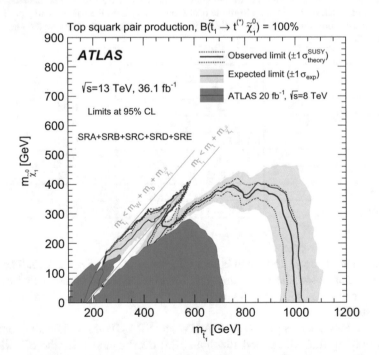

Fig. 11.42 Observed (red solid line) and expected (blue solid line) exclusion contours at 95% CL as a function of $m_{\tilde{t}_1}$ and $m_{\tilde{\chi}_1^0}$ in the scenario where both top squarks decay via $\tilde{t}_1 \rightarrow t + \tilde{\chi}_1^0$. Masses that are within the contours are excluded. Uncertainty bands corresponding to the $\pm 1\sigma$ variation of the expected limit (yellow band) and the sensitivity of the observed limit to $\pm 1\sigma$ variations of the signal theoretical uncertainties (red dotted lines) are also indicated. Observed limits from all third-generation Run-1 searches (cf. Fig. 11.2) at $\sqrt{s} = 8$ TeV overlaid for comparison in blue [22]

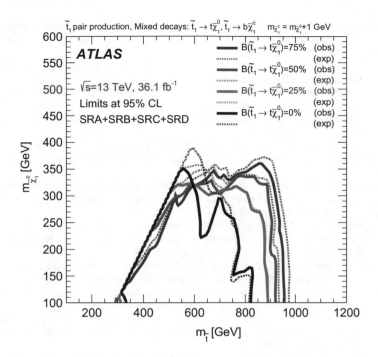

Fig. 11.43 Observed (solid line) and expected (dashed line) exclusion contours at 95% CL as a function of $m_{\tilde{t}_1}$ and $m_{\tilde{\chi}_1^0}$ and $\mathcal{BR}(\tilde{t}_1 \to t + \tilde{\chi}_1^0)$ in the natural SUSY-inspired mixed grid scenario where $m_{\tilde{\chi}_1^\pm} = m_{\tilde{\chi}_1^0} + 1\,\mathrm{GeV}$ [22]

squark masses up to 1 TeV can be excluded for neutralino masses up to 300 GeV. In the intermediate top squark mass range, neutralino masses up to 400 GeV can be excluded.

Figure 11.43 shows the exclusion contour at 95% CL obtained by combining all SRs as a function of the top squark and neutralino masses depending on $\mathcal{BR}(\tilde{t}_1 \to t + \tilde{\chi}_1^0)$. For pure $\tilde{t}_1 \to b + \tilde{\chi}_1^\pm$ decays, top squark masses up to 800 GeV can be excluded for neutralino masses up to 100 GeV while neutralino masses up to 350 GeV can be excluded for a top squark mass of 550 GeV.

References

1. Collaboration ATLAS (2014) Search for direct pair production of the top squark in all-hadronic final states in proton-proton collisions at $\sqrt{s} = 8$ TeV with the ATLAS detector. JHEP 09:015. https://doi.org/10.1007/JHEP09(2014)015
2. Collaboration ATLAS (2014) Search for top squark pair production in final states with one isolated lepton, jets, and missing transverse momentum in $\sqrt{s} = 8$ TeV pp collisions with the ATLAS detector. JHEP 11:118. https://doi.org/10.1007/JHEP11(2014)118

3. Collaboration ATLAS (2014) Search for direct top-squark pair production in final states with two leptons in pp collisions at $\sqrt{s} = 8$ TeV with the ATLAS detector. JHEP 06:124. https://doi.org/10.1007/JHEP06(2014)124
4. Collaboration ATLAS (2015) Measurement of Spin correlation in Top-Antitop quark events and search for top squark pair production in pp collisions at $\sqrt{s} = 8$ TeV using the ATLAS detector. Phys Rev Lett 114:142001. https://doi.org/10.1103/PhysRevLett.114.142001
5. Collaboration ATLAS (2015) ATLAS Run 1 searches for direct pair production of third-generation squarks at the Large Hadron Collider. Eur Phys J C 75:510. https://doi.org/10.1140/epjc/s10052-015-3726-9
6. Collaboration CMS (2015) Search for supersymmetry using razor variables in events with b-tagged jets in pp collisions at $\sqrt{s} = 8$ TeV. Phys Rev D 91:052018. https://doi.org/10.1103/PhysRevD.91.052018
7. Collaboration CMS (2017) Search for top squark pair production in compressedmass-spectrum scenarios in proton-proton collisions at $\sqrt{s} = 8$ TeV using the α_T variable. Phys Lett B 767:403. https://doi.org/10.1016/j.physletb.2017.02.007
8. Collaboration CMS (2014) Search for pair production of third-generation scalar leptoquarks and top squarks in proton-proton collisions at $\sqrt{s} = 8$ TeV. Phys Lett B 739:229. https://doi.org/10.1016/j.physletb.2014.10.063. arXiv:1408.0806 [hep-ex]
9. Alwall J, Schuster P, Toro N (2009) Simplified models for a first characterization of new physics at the LHC. Phys Rev D 79:075020. https://doi.org/10.1103/PhysRevD.79.075020
10. Alves D (2012) Simplified models for LHC new physics searches. J Phys G 39. Arkani-Hamed N et al (ed), p 105005. https://doi.org/10.1088/0954-3899/39/10/105005
11. Alwall J et al (2014) The automated computation of tree-level and next-to-leading order differential cross sections, and their matching to parton shower simulations. JHEP 07:079. https://doi.org/10.1007/JHEP07(2014)079
12. Lange DJ (2001) The EvtGen particle decay simulation package. Nucl Instrum Methods 462(1–2):152–155. https://doi.org/10.1016/S0168-9002(01)00089-4
13. Lönnblad L, Prestel S (2013) Merging multi-leg NLO matrix elements with parton showers. JHEP 03:166. https://doi.org/10.1007/JHEP03(2013)166
14. Ball Richard D et al (2013) Parton distributions with LHC data. Nucl Phys B 867:244–289. https://doi.org/10.1016/j.nuclphysb.2012.10.003
15. Beenakker W et al (1998) Stop production at hadron colliders. Nucl Phys B 515:3–14. https://doi.org/10.1016/S0550-3213(98)00014-5
16. Beenakker W et al (2010) Supersymmetric top and bottom squark production at hadron colliders. JHEP 08:098. https://doi.org/10.1007/JHEP08(2010)098
17. Beenakker W et al (2011) Squark and gluino hadroproduction. Int J Mod Phys A 26:2637–2664. https://doi.org/10.1142/S0217751X11053560
18. Borschensky C et al (2014) Squark and gluino production cross sections in pp collisions at $\sqrt{s} = 13$, 14, 33 and 100 TeV. Eur Phys J C 74:3174. https://doi.org/10.1140/epjc/s10052-014-3174-y
19. ATLAS Collaboration (2010) The simulation principle and performance of the ATLAS fast calorimeter simulation FastCaloSim. ATL-PHYS-PUB-2010-013. https://cds.cern.ch/record/1300517
20. ATLAS Collaboration (2016) Electron efficiency measurements with the ATLAS detector using the 2015 LHC proton-proton collision data. ATLAS-CONF-2016-024. 2016. https://cds.cern.ch/record/2157687
21. Collaboration ATLAS (2016) Muon reconstruction performance of the ATLAS detector in proton-proton collision data at $\sqrt{s} = 13$ TeV. Eur Phys J C 76:292
22. Collaboration ATLAS (2017) Search for a scalar partner of the top quark in the jets plus missing transverse momentum final state at $\sqrt{s} = 13$ TeV with the ATLAS detector. JHEP 12:085 arXiv:1709.04183 [hep-ex]
23. Collaboration ATLAS (2014) Electron and photon energy calibration with the ATLAS detector using LHC Run 1 data. Eur Phys J C 74:3071. https://doi.org/10.1140/epjc/s10052-014-3071-4

24. Collaboration ATLAS (2017) Performance of the ATLAS trigger system in 2015. Eur Phys J C 77:317. https://doi.org/10.1140/epjc/s10052-017-4852-3

25. Olive KA et al (2014) Rev Part Phys Chin Phys C38:090001. https://doi.org/10.1088/1674-1137/38/9/090001

26. Wasserstein Ronald L, Lazar Nicole A (2016) The ASA's statement on p-values: context, process, and purpose. Am Stat 70(2):129–133. https://doi.org/10.1080/00031305.2016.1154108

27. Brun R, Rademakers F (1997) ROOT-an object oriented data analysis framework. Nucl Instrum Methods Phys Res Sect A: Accel Spectrom Detect Assoc Equip 389:81–86

28. Lester CG, Summers DJ (1999) Measuring masses of semiinvisibly decaying particles pair produced at hadron colliders. Phys Lett B 463:99–103. https://doi.org/10.1016/S0370-2693(99)00945-4

29. Barr A, Lester CG, Stephens P (2003) A variable for measuring masses at hadron colliders when missing energy is expected; m T2: the truth behind the glamour. Nucl Part Phys 29:2343

30. Jackson P, Rogan C, Santoni M (2017) Sparticles in motion: analyzing compressed SUSY scenarios with a new method of event reconstruction. Phys Rev D 95(3):035031. https://doi.org/10.1103/PhysRevD.95.035031

31. Jackson P, Rogan C (2017) Recursive Jigsaw reconstruction: HEP event analysis in the presence of kinematic and combinatoric ambiguities. Phys Rev D 96(11):112007. https://doi.org/10.1103/PhysRevD.96.112007

32. An H, Wang L-T (2015) Opening up the compressed region of top squark searches at 13 TeV LHC. Phys Rev Lett 115:181602. https://doi.org/10.1103/PhysRevLett.115.181602

33. Macaluso S et al (2016) Revealing compressed stops using high-momentum recoils. JHEP 03:151. https://doi.org/10.1007/JHEP03(2016)151

34. Chamseddine AH, Arnowitt RL, Nath P (1982) Locally supersymmetric grand unification. Phys Rev Lett 49:970. https://doi.org/10.1103/PhysRevLett.49.970

35. Barbieri R, Ferrara S, Savoy CA (1982) Gauge models with Spontaneously Broken local supersymmetry. Phys Lett B 119:343. https://doi.org/10.1016/0370-2693(82)90685-2

36. Kane Gordon L et al (1994) Study of constrained minimal supersymmetry. Phys Rev D 49:6173–6210. https://doi.org/10.1103/PhysRevD.49.6173

37. Collaboration ATLAS (2016) Search for top squarks in final states with one isolated lepton, jets, and missing transverse momentum in $\sqrt{s} = 13$ TeV pp collisions with the ATLAS detector. Phys Rev D 94:052009. https://doi.org/10.1103/PhysRevD.94.052009

38. Fletcher GT (2015) Multijet background estimation ForSUSYSearches and particle flow offline reconstruction using the ATLAS Detector at the LHC. PhD thesis. SheffieldU, Mar 2015. http://inspirehep.net/record/1429579/files/fulltext_1Du5ll.pdf

39. Aad et al G (2009) Expected performance of the ATLAS experiment—detector, trigger and physics (2009). arXiv:0901.0512 [hep-ex]

40. ATLAS Collaboration (2013) Search for squarks and gluinos with the ATLAS detector in final states with jets and missing transverse momentum using 4.7fb^{-1} of $\sqrt{s} = 7$ TeV proton-proton collision data. Phys Rev D 87:012008. https://doi.org/10.1103/PhysRevD.87.012008

41. Schreyer M, Redelbach A, Ströhmer R (2015) Search for supersymmetry in events containing light leptons, jets and missing transverse momentum in $\sqrt{s} = 8$ TeV pp collisions with the ATLAS detector. Presented 25 Sep 2015. June 2015. https://cds.cern.ch/record/2055513

42. Nachman B, Lester CG (2013) Significance Variables. Phys Rev D88.7:075013. https://doi.org/10.1103/PhysRevD.88.075013

43. Collaboration ATLAS (2015) Jet energy measurement and its systematic uncertainty in proton-proton collisions at $\sqrt{s} = 7$ TeV with the ATLAS detector. Eur Phys J C 75:17. https://doi.org/10.1140/epjc/s10052-014-3190-y

44. Collaboration ATLAS (2013) Jet energy resolution in proton-proton collisions at $\sqrt{s} = 7$ TeV recorded in, (2010) with the ATLAS detector. Eur. Phys J C 73:2306. https://doi.org/10.1140/epjc/s10052-013-2306-0

45. Collaboration ATLAS (2016) Performance of b-jet identification in the ATLAS experiment. JINST 11:P04008. https://doi.org/10.1088/1748-0221/11/04/P04008

46. ATLAS Collaboration (2016) Optimisation of the ATLAS b-tagging performance for the 2016 LHC run. ATL-PHYS-PUB-2016-012. https://cds.cern.ch/record/2160731
47. Collaboration ATLAS (2017) Reconstruction of primary vertices at the ATLAS experiment in run 1 proton-proton collisions at the LHC. Eur Phys J C 77:332. https://doi.org/10.1140/epjc/s10052-017-4887-5
48. ATLAS Collaboration (2018) Performance of missing transverse momentum reconstruction with the ATLAS detector using proton-proton collisions at $\sqrt{s} = 13$ TeV. arXiv:1802.08168 [hep-ex]
49. ATLAS Collaboration (2013) Pile-up subtraction and suppression for jets in ATLAS. ATLAS-CONF-2013-083. https://cds.cern.ch/record/1570994
50. ATLAS Collaboration (2014) Tagging and suppression of pileup jets with the ATLAS detector. ATLAS-CONF-2014-018. https://cds.cern.ch/record/1700870
51. ATLAS Collaboration (2016) Measurement of the photon identification efficiencies with the ATLAS detector using LHC Run-1 data. arXiv:1606.01813 [hep-ex]
52. Gleisberg et al T (2009) Event generation with SHERPA 1.1. JHEP 02:007. https://doi.org/10.1088/1126-6708/2009/02/007
53. Bahr M et al (2008) Herwig++ physics and manual. Eur Phys J C 58:639–707. https://doi.org/10.1140/epjc/s10052-008-0798-9
54. Mrenna S, Richardson P (2004) Matching matrix elements and parton showers with HERWIG and PYTHIA. JHEP 05:040. https://doi.org/10.1088/1126-6708/2004/05/040. arXiv:hep-ph/0312274 [hep-ph]
55. Alwall J et al (2007) A standard format for Les Houches event files. Comput Phys Commun 176:300–304. https://doi.org/10.1016/j.cpc.2006.11.010
56. Gioacchino R (2012) The Profile likelihood ratio and the look elsewhere effect in high energy physics. Nucl Instrum Methods A661:77–85. https://doi.org/10.1016/j.nima.2011.09.047
57. Verkerke W, Kirkby DP (2003) The RooFit toolkit for data modeling. eConf C0303241:MOLT007. [physics]
58. Moneta L et al (2010) The roostats project. PoS ACAT2010 :057. arXiv:1009.1003 [physics.data-an]
59. Cranmer K et al (2012) HistFactory: a tool for creating statistical models for use with RooFit and RooStats
60. Baak M et al (2014) HistFitter software framework for statistical data analysis. arXiv:1410.1280 [hep-ex]
61. ATLAS Collaboration (2011) Procedure for the LHC Higgs boson search combination in summer 2011. ATL-PHYS-PUB-2011-011. https://cds.cern.ch/record/1375842
62. Cowan G et al (2011) Asymptotic formulae for likelihood-based tests of new physics. Eur Phys J C 71:1554. https://doi.org/10.1140/epjc/s10052-011-1554-0
63. Junk T (1999) Confidence level computation for combining searches with small statistics. Nucl Instrum Methods A434:435–443. https://doi.org/10.1016/S0168-9002(99)00498-2
64. Read AL (2002) Presentation of search results: the CL(s) technique. J Phys G28:2693–2704. https://doi.org/10.1088/0954-3899/28/10/313
65. Read AL (2000) Modified frequentist analysis of search results (The CL(s) method). In: Proceedings of the workshop on confidence limits, CERN, Geneva, Switzerland, 17-18 Jan 2000, pp. 81–101. http://weblib.cern.ch/abstract?CERN-OPEN-2000-205

Chapter 12
The Search for Dark Matter

Astrophysical observations have provided compelling arguments that 26% of the matter density of the universe is made of non-baryonic DM (cf. Chap. 3). In order to search for DM at the LHC without assuming the existence of SUSY, several searches for different DM production processes and decays are performed by both ATLAS and CMS. As discussed in Sect. 3.3.1, WIMPs are currently the most promising DM candidates. However, the detection of a single WIMP pair which is potentially produced in any process beyond the SM at the LHC (cf. Fig. 12.1a) is impossible since the experimental signature would be missing transverse energy only and no E_T^{miss} trigger would ensure the recording of the collision, since no E_T^{miss} can be computed in the case of no particles arising from the hard scattering of the collision. The common signature for DM searches at the LHC is the production of a WIMP pair in association with SM particles which makes it possible to trigger the data acquisition. This signature is shown schematically in Fig. 12.1b.

The simplest theoretical approach for the pair production of WIMPs would be the assumption of an Effective Field Theory (EFT) which covers the interaction in which the WIMPs and the associated SM particles are produced. However, the description by EFTs is only justified whenever there is a clear separation between the energy scale of the process to describe and the energy scale of the actual interaction in the experiment. This leads to theoretical shortcomings in terms of renormalizability when applying EFTs for describing production processes at LHC Run 2 energies [1–3].

One possible solution is the usage of simplified models as it is done for the simulation of the pair production of potential supersymmetric particles. A simplified model where a spin-0 *mediator* between the DM and SM sector is added is used to simulate the production of WIMP pairs in association with other SM particles [4]. The interaction between SM particles and WIMPs is resolved by s- or t-channel exchanges of the mediator in order to correctly account for off-shell and on-shell effects in WIMP pair production. By specifying the quantum numbers of both WIMPs and the mediator as well as requiring the interactions to be minimally flavour violating [5], the parameter spaces of the theoretical models remain low dimensional which allows for a simpler theoretical interpretation of experimental results [6]. Furthermore, in

© Springer Nature Switzerland AG 2019
N. M. Köhler, *Searches for the Supersymmetric Partner of the Top Quark, Dark Matter and Dark Energy at the ATLAS Experiment*,
Springer Theses, https://doi.org/10.1007/978-3-030-25988-4_12

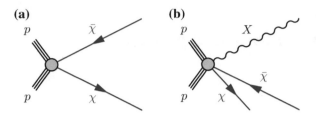

Fig. 12.1 Sketch of WIMP pair production with no other SM particle (**a**) and with associated production of a SM particle X (**b**) at the LHC

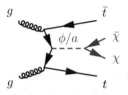

Fig. 12.2 Feynman diagram of gluon fusion producing a scalar (ϕ) or pseudoscalar (a) mediator in association with a top quark pair. The mediator subsequently decays into a pair of WIMPs

minimal flavour violating scenarios, the coupling strength of the spin-0 mediator, which can be scalar (ϕ) or pseudoscalar (a), is proportional to the Yukawa couplings of the fermions [7]. Thus, the highest production rates at LHC collisions are expected for couplings of the mediator to top quarks. The WIMP is assumed to be a Dirac fermion.

Figure 12.2 shows the Feynman diagram of gluon fusion producing a spin-0 mediator in association with a top quark pair which subsequently decays into a pair of WIMPs. For fully hadronic top quark decays, the experimental signature is missing transverse energy and the presence of multiple jets.

12.1   Simulation of WIMP Pair Production

The simplified model used for describing the production of the spin-0 mediator in association with a top quark pair and subsequent decay into a pair of WIMPs, has four free parameters: the mediator mass m_ϕ or m_a, the WIMP mass m_χ and the two coupling constants of the mediator to the WIMP and the top quark, g_χ and g_t, respectively [7]. Further assumptions are that the width of the mediator calculated from the four parameters of the model is minimal, the branching fraction of the mediator decay to WIMPs is $\mathcal{BR}(\chi/a \to \chi\bar{\chi}) = 100\%$ and the mediator couplings to WIMPs and top quarks are of equal size, $g_\chi = g_t$ [4]. Events were generated from LO matrix elements using MADGRAPH 2.3.3 and the parton showering was performed in PYTHIA 8.212 with the A14 underlying-event tune. NNPDF2.3LO was used as the PDF set and the jet-parton matching was done following the CKKW-L prescription with the matching scale being set to one quarter of the mediator mass. The signal cross sections are calculated to NLO in the strong coupling constant using the MADGRAPH5_aMC@NLO generator with the NNPDF3.0NNLO PDF set.

As for the simulation of top squark pair production, the simulated signals are processed through a fast simulation framework where the showers in the electromagnetic and hadronic calorimeters are simulated with a parametrised description. The fast simulation framework was validated against the full GEANT 4 simulation for several selected signal samples (cf. Appendix C.1) and subsequently used for all signal processes.

12.2   Basic Experimental Signature

The same lepton identification criteria as in the search for top squarks are used for defining the lepton veto. Figure 12.3a shows the relative number of jets after applying a lepton veto for three scenarios of WIMP pair production through different mediators as well for the sum of SM processes apart from multijet production. Requiring at least four jets retains more than 94% of potential DM events for all scenarios shown while more than 30% of the SM contributions are rejected. The relative number of b-jets depicted in Fig. 12.3b suggests that requiring at least two b-jets rejects more than 90% of the SM backgrounds but also roughly 50% of potential signal contributions. However, due to the enormous suppression factor against SM backgrounds, at least two b-jets are required for all SRs targeting WIMP pair production in association with fully-hadronically decaying top quark pairs. Figure 12.3b also shows that the relative number of b-jets is independent on whether the mediator is a scalar (ϕ) or a pseudoscalar (a).

The relative distributions of the transverse momenta of the second and fourth jet leading in p_T are shown in Fig. 12.4 for both the SM contributions and the three

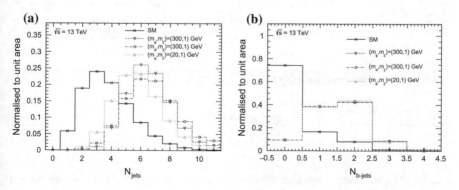

Fig. 12.3 Relative number of simulated jets and b-jets after applying a lepton veto and requiring the scalar sum of transverse momenta of all jets, H_T, to be $H_T > 150$ GeV. A selection of WIMP pair production processes as well as the sum of the SM contributions apart from multijet production are shown

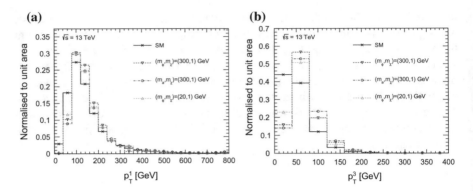

Fig. 12.4 Relative distributions of the jet transverse momenta p_T for the second and fourth jet leading in p_T after applying a lepton veto, $H_T > 150$ GeV, as well as at least four jets and two b-jets for the same selection of signal processes shown in Fig. 12.3. The distributions of the first and third jet leading in p_T can be found in Appendix C.2

Table 12.1 Preselection applied for the SRs defined in the search for the pair production of WIMPs in association with a top quark pair. $\mathscr{L}$ [10^{34} cm^{-2}s^{-1}] is denoting the instantaneous luminosity. A τ is defined as a non-b-jet within $|\eta| < 2.5$ with fewer than four associated charged-particle tracks with $p_T > 500$ MeV in case the angular separation between the jet and the $\mathbf{p}_T^{\mathrm{miss}}$ in the transverse plane is less than $\pi/5$

	Requirement
Number of leptons	$N_\ell = 0$
Trigger	E_T^{miss}, HLT threshold: 70 GeV (2015)
	E_T^{miss}, HLT threshold: 90 GeV (2016, $\mathscr{L} \leq 1.02$)
	E_T^{miss}, HLT threshold: 100 GeV (2016, $1.02 < \mathscr{L} \leq 1.13$)
	E_T^{miss}, HLT threshold: 110 GeV (2016, $\mathscr{L} > 1.13$)
Missing transverse energy	$E_T^{\mathrm{miss}} > 250$ GeV
Number of jets	$N_{\mathrm{jets}} \geq 4$
Jet transverse momenta (sorted by p_T)	>80 GeV, >80 GeV, >40 GeV, >40 GeV
Number of b-jets	$N_{b-\mathrm{jets}} \geq 2$
$\left|\Delta\phi\left(\mathrm{jet}^{0-3}, E_T^{\mathrm{miss}}\right)\right|$	>0.4
Track-based missing transverse energy $E_T^{\mathrm{miss,track}}$	>30 GeV
$\left|\Delta\phi\left(E_T^{\mathrm{miss}}, E_T^{\mathrm{miss,track}}\right)\right|$	$< \frac{\pi}{3}$
τ-veto	$N_\tau = 0$

simulated signal scenarios which are also shown in Fig. 12.3. As in the search for top squark pair production, requiring the first two jets leading in p_T to fulfil $p_T > 80$ GeV rejects 10% of the signal contributions, but almost the double amount of the SM contributions (cf. Fig. 12.4a). Figure 12.4b indicates that requiring the fourth jet leading in p_T to fulfil $p_T > 40\,\text{GeV}$, roughly 45% of the SM contributions can be suppressed by only rejecting between 10 and 20% of potential signal contributions depending on the choice of the model parameters.

The striking similarities in the basic requirements of the search for WIMPS compared to the search for top squarks (cf. Chap. 11) result in the circumstance that an almost identical basic selection as summarised in Table 11.1 is exploited for this search. While the search for top squarks only requires the presence of at least two b-jets in the non-compressed SRs, for this search, at least two b-jets are required throughout all SRs and CRs. To suppress contributions from multi-jet production, the jet smearing technique applied to the SRs defined in the following revealed that a better suppression of the multi-jet background can be achieved when using the four highest-p_T jets for calculating the minimum difference in the azimuthal angle between the jet and E_T^{miss}, $\left| \Delta\phi\left(\text{jet}^{0-3}, E_T^{\text{miss}} \right) \right|$. This can be explained by the fact that the final requirement on the missing transverse energy are softer than in the search for top squarks. The common basic requirements used for the search for DM are summarised in Table 12.1.

12.3 Signal Region Definitions

In order to study differences in the decay topologies between top squarks and mediator-WIMP production, the relative distributions of the missing transverse energy and the transverse energy of the leading p_T jet are shown in Fig. 12.5. WIMP pair production results in slightly softer kinematic spectra than predicted for top squark decays (cf. Fig. 12.5a). Differences in the mediator mass have smaller impact on the kinematic spectra than the difference arising from the underlying theoretical model since for top squarks, both the neutralino and the top quark are originating from the top squark whereas in the case of WIMP production, two top quarks and the mediator have to be produced in the hard process which diminishes the available energy for the WIMPs. Thus, for the search for WIMPs, two new SRs are constructed targeting small and large mediator masses, respectively.

Low Mediator Masses

For the optimisation of the SR targeting low mediator masses simulated events with a scalar mediator of mass $m_\phi = 20\,\text{GeV}$ and a WIMP mass of $m_\chi = 1\,\text{GeV}$ are exploited as the benchmark scenario. The coupling between mediator and SM or mediator and WIMP, respectively, is set to $g = g_\chi = g_t = 1$. Since WIMP signatures can potentially lead to softer kinematic distributions, no requirement on $m_T^{b,\text{min}}$ is applied for the optimisation. Looking at the simulated distribution of the distance between the two b-jets with the highest MV2 discriminant, $\Delta R\,(b, b)$ (cf. Fig. 12.6a), requiring $\Delta R\,(b, b) > 1.5$ leads to the maximum expected significance for the benchmark scenario. The simulated distribution of $m_T^{b,\text{min}}$ with the expected significances depending on a potential cut on $m_T^{b,\text{min}}$ are depicted in Fig. 12.6b. While for the benchmark scenario, the maximum expected significance is reached for requiring

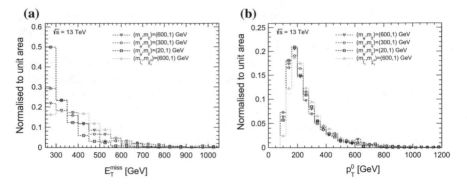

Fig. 12.5 Relative distributions of the missing transverse energy (**a**) and the transverse momentum of the jet leading in p_T (**b**) for three simulated Dark Matter (blue, red and black) and one SUSY (green) scenario. The rightmost bin includes overflow events

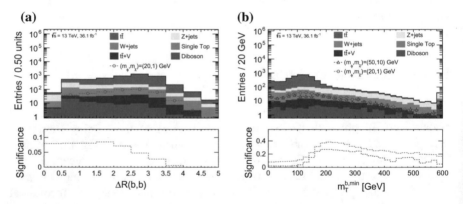

Fig. 12.6 Simulated distributions of $\Delta R\,(b, b)$ (**a**) and $m_T^{b,\min}$ (**b**) after applying the requirements summarised in Table 12.1 for the SM processes and signal scenarios with $(m_\phi, m_\chi = (20, 1)$ GeV and $(m_\phi, m_\chi = 50, 10)$ GeV, respectively, The bottom panel shows the expected significances of the signal scenario depending on a potential cut value on the corresponding variable

$m_T^{b,\min} > 200$ GeV, looking at a scenario with $m_\phi = 50$ GeV and $m_\chi = 10$ GeV, the maximum value of the expected significance is shifted to lower values. To retain the significance for similar signal scenarios, a requirement of $m_T^{b,\min} > 150$ GeV is chosen.

Figure 12.7a shows the simulated distribution of $m_T^{b,\max}$ after the requirements on $\Delta R\,(b, b)$ and $m_T^{b,\min}$ are applied. The maximum expected significance is reached for $m_T^{b,\max} > 250$ GeV. The expected significance depending on the second leading $R = 0.8$ reclustered jet mass, $m_{\text{jet}, R=0.8}^1$ reaches a plateau at roughly 80 GeV but has its global maximum at $m_{\text{jet}, R=0.8}^1 > 150$ GeV. However, requiring $m_{\text{jet}, R=0.8}^1 > 150$ GeV would result in a huge statistical uncertainty of the signal simulation and hence $m_{\text{jet}, R=0.8}^1 > 80$ GeV is chosen.

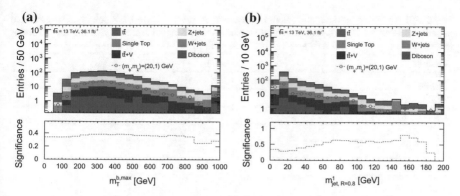

Fig. 12.7 Simulated distributions of $m_\mathrm{T}^{b,\max}$ (**a**) and $m_{\mathrm{jet},R=0.8}^1$ (**b**) with the expected significance of the signal scenario with $(m_\phi, m_\chi = (20, 1)\,\mathrm{GeV}$ depending on a potential cut value on the corresponding variable shown in the bottom panel. In addition to the basic requirements summarised in Table 12.1, $\Delta R\,(b, b) > 1.5$ and $m_\mathrm{T}^{b,\min} > 200\,\mathrm{GeV}$ are required

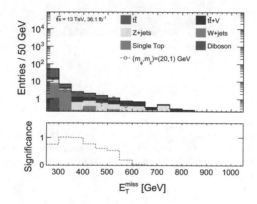

Fig. 12.8 Simulated distribution of $E_\mathrm{T}^\mathrm{miss}$ with the expected significance of the signal scenarios with $m_\phi = 20\,\mathrm{GeV}$ and $m_\chi = 1\,\mathrm{GeV}$ depending on a potential cut value on the corresponding variable shown in the bottom panel. All requirements from SRt1 (cf. Table 12.2) except for the one on $E_\mathrm{T}^\mathrm{miss}$ are applied

Figure 12.8 shows the simulated distribution of $E_\mathrm{T}^\mathrm{miss}$ after applying all requirements of the SR targeting low mediator masses except for the one on $E_\mathrm{T}^\mathrm{miss}$. The expected significance reaches its maximum for requiring $E_\mathrm{T}^\mathrm{miss} > 300\,\mathrm{GeV}$. All selection criteria for the SR targeting low mediator masses are summarised in Table 12.2.

High Mediator Masses

For the optimisation of the SR targeting high mediator masses simulated events with a scalar mediator of mass $m_\phi = 300\,\mathrm{GeV}$ and a WIMP mass of $m_\chi = 1\,\mathrm{GeV}$ are exploited. The coupling between mediator and SM or mediator and WIMP, respectively, is set to $g = g_\chi = g_t = 2$. The $m_\mathrm{T}^{b,\min}$ requirement is tightened to

Table 12.2 Summary of the requirements to be made in addition to the basic selection summarised in Table 12.1 for the SR targeting high mediator masses. The signal region is optimised for a scalar mediator of mass $m_\phi = 20\,\text{GeV}$ and a WIMP mass of $m_\chi = 1\,\text{GeV}$ with a coupling $g = 1$

Variable	SRt1
$m^0_{\text{jet}, R=0.8}$	$>80\,\text{GeV}$
$m^1_{\text{jet}, R=0.8}$	$>80\,\text{GeV}$
$m^{b,\text{max}}_{\text{T}}$	$>250\,\text{GeV}$
$m^{b,\text{min}}_{\text{T}}$	$>150\,\text{GeV}$
$E^{\text{miss}}_{\text{T}}$	$>300\,\text{GeV}$
$\Delta R\,(b, b)$	>1.5

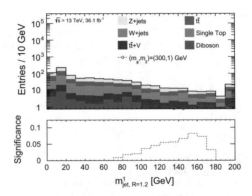

Fig. 12.9 Simulated distribution of $m^1_{\text{jet}, R=1.2}$ with the expected significance of the signal scenarios with $m_\phi = 300\,\text{GeV}$ and $m_\chi = 1\,\text{GeV}$ depending on a potential cut value on the corresponding variable shown in the bottom panel. In addition to the basic requirements summarised in Table 12.1, $m^{b,\text{min}}_{\text{T}} > 200\,\text{GeV}$ is required

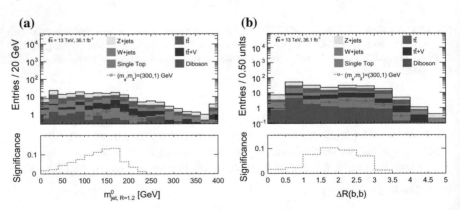

Fig. 12.10 Simulated distributions of $m^0_{\text{jet}, R=1.2}$ (**a**) and $\Delta R\,(b, b)$ (**b**) with the expected significance of the signal scenarios with $m_\phi = 300\,\text{GeV}$ and $m_\chi = 1\,\text{GeV}$ depending on a potential cut value on the corresponding variable shown in the bottom panel. In addition to the basic requirements summarised in Table 12.1, $m^{b,\text{min}}_{\text{T}} > 200\,\text{GeV}$ and $m^1_{\text{jet}, R=1.2} > 80\,\text{GeV}$ are required

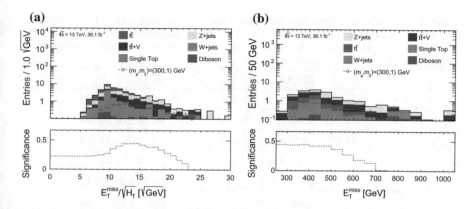

Fig. 12.11 Simulated distributions of $E_T^{miss}/\sqrt{H_T}$ (a) and E_T^{miss} (b) with the expected significance of the signal scenarios with $m_\phi = 300\,\text{GeV}$ and $m_\chi = 1\,\text{GeV}$ depending on a potential cut value on the corresponding variable shown in the bottom panel. For **a** all requirements from SRt2 (cf. Table 12.3) except for the ones on $E_T^{miss}/\sqrt{H_T}$ and E_T^{miss} are applied, for **b** all requirements from SRt1 except for the one on E_T^{miss} are applied

Table 12.3 Summary of the requirements to be made in addition to the basic selection summarised in Table 12.1 for the SR targeting high mediator masses. The signal region is optimised for a scalar mediator of mass $m_\phi = 300\,\text{GeV}$ and a WIMP mass of $m_\chi = 1\,\text{GeV}$ with a coupling $g = 2$

Variable	SRt2
$m_{jet, R=1.2}^0$	$>140\,\text{GeV}$
$m_{jet, R=1.2}^1$	$>80\,\text{GeV}$
$m_T^{b, min}$	$>200\,\text{GeV}$
$E_T^{miss}/\sqrt{H_T}$	$>12\sqrt{\text{GeV}}$
E_T^{miss}	$>300\,\text{GeV}$
$\Delta R(b, b)$	>1.5

$m_T^{b, min} > 200\,\text{GeV}$ in addition to the basic preselection summarised in Table 12.1. Figure 12.9 shows the simulated distribution of the second leading $R = 1.2$ reclustered jet mass applying the aforementioned selection. Requiring $m_{jet, R=1.2}^1 > 80\,\text{GeV}$ is suppressing most of the SM contributions and has no impact on the expected significance. This requirement applied, the expected significances depending on the requirements made on $m_{jet, R=1.2}^0$ (cf. Fig. 12.10(a)) and $\Delta R(b, b)$ (cf. Fig. 12.10(b)) are maximum in case of requiring $m_{jet, R=1.2}^0 > 140\,\text{GeV}$ and $\Delta R(b, b) > 1.5$.

After all requirements introduced so far for the SR targeting large mediator masses have been applied, the expected significance depending on the cut value on $E_T^{miss}/\sqrt{H_T}$, where H_T is the scalar sum of transverse momenta of all jets, is shown in Fig. 12.11a. The maximum expected sensitivity is reached when requiring $E_T^{miss}/\sqrt{H_T} > 12\sqrt{\text{GeV}}$. All requirements made for the SR targeting high mediator masses are summarised in Table 12.3.

Table 12.4 Expected number of simulated events in the signal regions targeting WIMP pair production scaled to an integrated luminosity of 36.1 fb^{-1}. For the signal scenarios used for the optimisation of the SRs, the expected significance is also given. For the SM processes, the statistical and the experimental systematic uncertainties are shown

	SRt1	SRt2
$m_\phi = 300\,\text{GeV}, m_\chi = 1\,\text{GeV}$	$2.76 \pm 0.15\ (0.2\sigma)$	$4.66 \pm 0.18\ (0.4\sigma)$
$m_a = 300\,\text{GeV}, m_\chi = 1\,\text{GeV}$	$3.06 \pm 0.34\ (0.2\sigma)$	$6.20 \pm 0.55\ (0.6\sigma)$
$m_\phi = 20\,\text{GeV}, m_\chi = 1\,\text{GeV}$	$9.33 \pm 1.03\ (1.0\sigma)$	$12.82 \pm 1.18\ (1.4\sigma)$
$m_a = 20\,\text{GeV}, m_\chi = 1\,\text{GeV}$	$7.62 \pm 0.39\ (0.8\sigma)$	$12.11 \pm 0.45\ (1.3\sigma)$
$t\bar{t}$	$6.12 \pm 0.50^{+1.376}_{-1.105}$	$2.79 \pm 0.30^{+0.559}_{-0.630}$
Z+jets	$3.24 \pm 0.34^{+0.717}_{-0.991}$	$5.72 \pm 0.72^{+1.226}_{-1.182}$
W+jets	$3.31 \pm 2.67^{+2.990}_{-3.071}$	$1.28 \pm 0.45^{+0.291}_{-0.218}$
Single Top	$1.33 \pm 0.21^{+0.276}_{-0.266}$	$1.99 \pm 0.30^{+0.307}_{-0.645}$
$t\bar{t} + V$	$3.53 \pm 0.34^{+0.382}_{-0.345}$	$5.60 \pm 0.44^{+0.809}_{-0.579}$
Diboson	$0.45 \pm 0.15^{+0.196}_{-0.224}$	$0.86 \pm 0.24^{+0.214}_{-0.216}$
Total SM	$17.97 \pm 2.77^{+3.246}_{-3.399}$	$18.24 \pm 1.07^{+2.698}_{-2.631}$

12.4 Background Estimation

Comparing the SM contributions in SRt1 and SRt2 with the SRs targeting top squark pair production, the dominating SM contributions are almost identical due to the similarities in the SR requirements (cf. Table 12.4). Consequently, in order to estimate the normalisation of the SM processes in the SRs, similar approaches as in the search for top squark pair production is employed. The normalisation of SM $t\bar{t}$ production is estimated using CRs with exactly one lepton. The normalisation of the Z+jets contribution is estimated in a two lepton CR while the normalisation of the $t\bar{t} + V$ processes is estimated using a $t\bar{t} + \gamma$ CR. The SR contributions of single top quark as well as diboson production are less than 11 and 5% respectively. Thus, their normalisation is entirely taken from simulation. The same is done for W+jets production although its contribution in SRt1 is more than 18% but only 7% in SRt2.

Top Quark Pair Production

For defining a CR enriching SM $t\bar{t}$ production, the same strategy as described in Sect. 11.4 is followed. Exactly one lepton with $30\,\text{GeV} < m_\text{T}\left(\ell, E_\text{T}^\text{miss}\right) < 100\,\text{GeV}$ is required accounting for a lepton which was not reconstructed in case semi-leptonic $t\bar{t}$ decays are contributing in the fully-hadronic final state. All basic requirements summarised in Table 12.1 except for the lepton veto are applied but the lepton is treated as a non-b-tagged jet for the requirements on the jet multiplicity and transverse momenta as described in Sect. 11.4. In order to define CRs targeting SRt1 and SRt2, the requirement on $m_\text{T}^{b,\text{min}}$ is loosened to $m_\text{T}^{b,\text{min}} > 100\,\text{GeV}$ in order to enrich the $t\bar{t}$

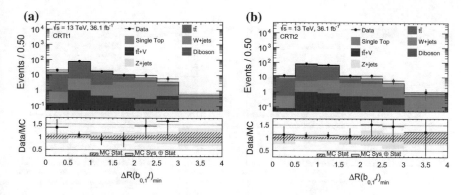

Fig. 12.12 Comparison of data and simulated SM contributions for the $\Delta R(b_{0,1}, \ell)_{\min}$ distribution after applying all requirements from CRTt1 (**a**) and CRTt2 (**b**) except for the one on $\Delta R(b_{0,1}, \ell)_{\min}$. The yellow bands show the quadratic sum of the statistical and experimental systematic uncertainties

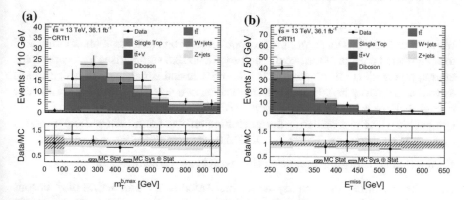

Fig. 12.13 Comparison of data and simulated SM contributions for the distributions of $m_{\mathrm{T}}^{b,\max}$ (**a**) and $E_{\mathrm{T}}^{\mathrm{miss}}$ (**b**) after applying all requirements from CRTt1. The yellow bands show the quadratic sum of the statistical and experimental systematic uncertainties

contribution. The SRt1 requirements on $m_{\mathrm{T}}^{b,\max}$ and $E_{\mathrm{T}}^{\mathrm{miss}}$ are dropped and the requirements on the $R = 0.8$ reclustered jet masses are loosened to $m_{\mathrm{jet}, R=0.8}^{0} > 60\,\mathrm{GeV}$ and $m_{\mathrm{jet}, R=0.8}^{1} > 60\,\mathrm{GeV}$ in order to increase the statistics of the CR. Figure 12.12a shows the distribution of $\Delta R(b_{0,1}, \ell)_{\min}$ applying the requirements discussed above. Requiring $\Delta R(b_{0,1}, \ell)_{\min} < 1$ increases the purity of the $t\bar{t}$ contribution. Figure 12.13 shows the distributions of $m_{\mathrm{T}}^{b,\max}$ and $E_{\mathrm{T}}^{\mathrm{miss}}$ in the $t\bar{t}$ CR targeting SRt1. The shape of the distributions is flat which hints to the fact that dropping the requirements on these variables does not lead to a significant change of the event topology.

To define a $t\bar{t}$ CR targeting SRt2, the requirements on $E_{\mathrm{T}}^{\mathrm{miss}}/\sqrt{H_{\mathrm{T}}}$ and $E_{\mathrm{T}}^{\mathrm{miss}}$ are dropped. The resulting distribution of $\Delta R(b_{0,1}, \ell)_{\min}$ is shown in Fig. 12.12b. Requiring $\Delta R(b_{0,1}, \ell)_{\min} < 1.5$ increases the purity of the $t\bar{t}$ contribution in a region which is close in phase space to SRt2. The distributions of $E_{\mathrm{T}}^{\mathrm{miss}}/\sqrt{H_{\mathrm{T}}}$ and $E_{\mathrm{T}}^{\mathrm{miss}}$ in the $t\bar{t}$ CR targeting SRt2 are shown in Fig. 12.14. The shape of the distributions is flat

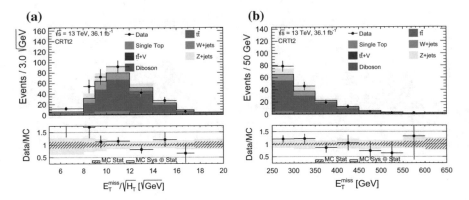

Fig. 12.14 Comparison of data and simulated SM contributions for the distributions of $E_T^{miss}/\sqrt{H_T}$ (**a**) and E_T^{miss} (**b**) after applying all requirements from CRTt2. The yellow bands show the quadratic sum of the statistical and experimental systematic uncertainties

which hints to the fact that dropping the requirements on these variables does not lead to a significant change of the event topology. Table 12.5 summarises the event yields after applying all $t\bar{t}$ CR requirements on both data and simulated SM backgrounds as well as the purity and the naive normalisation computed by using Eq. (11.5). The $t\bar{t}$ contributions in the CRs have a purities of 92-93% and the normalisation factor is compatible with one considering the quadratic sum of the statistical and experimental systematic uncertainties. The potential DM signal contamination is below 1%.

Production of Z Bosons in Association with Heavy-Flavour Jets

As in the search for top squarks, the normalisation of the production of Z bosons in association with heavy-flavour jets is estimated using a CR containing two charged leptons with same flavour but opposite charge and a transverse momentum of $p_T > 28$ GeV. The lowest unprescaled lepton triggers (cf. Table 11.9) have to trigger the data acquisition and the invariant mass of the leptons has to be within 5 GeV of the Z boson mass. Analogously to Sect. 11.4, $E_T^{miss} < 50$ GeV is required (cf. Fig. 11.34(a)) and the transverse momenta of the leptons are vectorially removed from the E_T^{miss} calculation resulting in $E_T^{miss'}$. In contrary to the search for top squarks, the requirement on $E_T^{miss'}$ is increased to $E_T^{miss'} > 160$ GeV since a larger cut value rather leads to a phase space more close to the SRs which have a $E_T^{miss} > 250$ GeV requirement. The same requirements on the jet multiplicity and transverse momenta are applied as in the SRs. In order to increase the statistics, for the Z+jets CR targeting SRt1, the requirements on $\Delta R\,(b, b)$, $m_T^{b,min}$ and $m_{jet,R=0.8}^1$ are dropped and the requirements on $m_T^{b,max}$ and $m_{jet,R=0.8}^0$ are loosened to $m_T^{b,max} > 100$ GeV and $m_{jet,R=0.8}^0 > 60$ GeV. For the CR targeting SRt2, the requirements on $\Delta R\,(b, b)$ and $m_{jet,R=1.2}^1$ are dropped and the requirements on $m_T^{b,min}$, $E_T^{miss}/\sqrt{H_T}$ and $m_{jet,R=1.2}^0$ are loosened to $m_T^{b,min} > 100$ GeV, $E_T^{miss}/\sqrt{H_T} > 6\sqrt{GeV}$ and $m_{jet,R=1.2}^0 > 60$ GeV.

Table 12.5 Composition of SM processes in $t\bar{t}$ control regions for 36.1 fb^{-1} of pp collision data. None of the simulated SM predictions is normalised using normalisation factors derived from a simultaneous fit (detailed in Sect. 11.6). The purities of the simulated $t\bar{t}$ contributions as well as the normalisation factors computed with Eq. (10.1) are also shown for all regions. For the simulated events, the statistical and the experimental systematic uncertainties are shown. The uncertainty on the purity and the normalisation factor includes the quadratic sum of the statistical and experimental systematic uncertainties

	CRTt1	CRTt2
$t\bar{t}$	$82.10 \pm 2.37^{+9.947}_{-8.845}$	$137.18 \pm 3.14^{+20.405}_{-19.485}$
Z+jets	$0.02 \pm 0.02 \pm 0.008$	$0.04 \pm 0.02 \pm 0.012$
W+jets	$1.44 \pm 0.54^{+0.592}_{-0.501}$	$4.15 \pm 0.98^{+0.993}_{-1.374}$
Single Top	$3.80 \pm 0.66^{+0.744}_{-0.730}$	$6.14 \pm 0.95^{+0.901}_{-0.942}$
$t\bar{t} + V$	$1.10 \pm 0.19^{+0.216}_{-0.213}$	$1.84 \pm 0.20^{+0.468}_{-0.455}$
Diboson	0.00 ± 0.00	0.00 ± 0.00
Total MC	$88.47 \pm 2.53^{+10.410}_{-9.032}$	$149.34 \pm 3.43^{+21.502}_{-20.084}$
Data	100.00 ± 10.00	164.00 ± 12.81
Purity of $t\bar{t}$ [%]	92.80 ± 11.55	91.86 ± 13.56
Normalisation of $t\bar{t}$	1.14 ± 0.19	1.11 ± 0.19

Table 12.6 Composition of SM processes in Z+jets control regions for 36.1 fb^{-1} of pp collision data. None of the simulated SM predictions is normalised using normalisation factors derived from a simultaneous fit (detailed in Sect. 11.6). The purities of the simulated Z+jets contributions as well as the normalisation factors computed with Eq. (10.1) are also shown for all regions. For the simulated events, the statistical and the experimental systematic uncertainties are shown. The uncertainty on the purity and the normalisation factor includes the quadratic sum of the statistical and experimental systematic uncertainties

	CRZt1	CRZt2
$t\bar{t}$	$1.24 \pm 0.56^{+0.990}_{-1.209}$	$1.74 \pm 0.66^{+0.932}_{-0.990}$
Z+jets	$90.93 \pm 1.94^{+15.265}_{-15.877}$	$110.29 \pm 2.18^{+17.666}_{-19.534}$
W+jets	0.00 ± 0.00	0.00 ± 0.00
Single Top	0.00 ± 0.00	$0.23 \pm 0.23^{+0.037}_{-0.036}$
$t\bar{t} + V$	$16.55 \pm 0.35^{+1.897}_{-1.767}$	$18.55 \pm 0.37^{+2.206}_{-2.063}$
Diboson	$9.95 \pm 1.93^{+2.831}_{-3.057}$	$10.73 \pm 2.40^{+2.446}_{-3.040}$
Total MC	$118.66 \pm 2.81^{+18.786}_{-19.059}$	$141.55 \pm 3.34^{+21.156}_{-22.184}$
Data	136.00 ± 11.66	155.00 ± 12.45
Purity of Z+jets [%]	76.62 ± 12.55	77.92 ± 12.45
Normalisation of Z+jets	1.19 ± 0.25	1.12 ± 0.23

Table 12.7 Composition of SM processes in the $t\bar{t} + \gamma$ control region for 36.1 fb^{-1} of pp collision data. None of the simulated SM predictions is normalised using normalisation factors derived from a simultaneous fit (detailed in Sect. 11.6). The purity of the simulated $t\bar{t} + \gamma$ contribution as well as the normalisation factor computed with Eq. (10.1) are also shown. For the simulated events, the statistical and the experimental systematic uncertainties are shown. The uncertainty on the purity and the normalisation factor includes the quadratic sum of the statistical and experimental systematic uncertainties

	CRγ
$t\bar{t}$	$11.33 \pm 1.75^{+2.411}_{-1.709}$
Z+jets	$0.56 \pm 0.16^{+0.133}_{-0.210}$
W+jets	$0.03 \pm 0.02^{+0.011}_{-0.016}$
Single Top	$1.86 \pm 0.77^{+0.458}_{-0.584}$
$t\bar{t} + V$	$1.77 \pm 0.24^{+0.323}_{-0.307}$
$t\bar{t} + \gamma$	$89.57 \pm 1.38^{+9.332}_{-9.133}$
$\gamma + V$	$4.99 \pm 0.60^{+1.123}_{-1.176}$
Diboson	0.00 ± 0.00
Total MC	$110.11 \pm 2.45^{+12.502}_{-11.522}$
Data	124.00 ± 11.14
Purity of $t\bar{t} + \gamma$ [%]	81.35 ± 9.49
Normalisation of $t\bar{t} + \gamma$	1.16 ± 0.18

Table 12.6 summarises the event yields after applying all Z+jets CR requirements on both data and simulated SM backgrounds as well as the purity and the naive normalisation computed by using Eq. (11.5). The Z+jets contributions in the CRs have purities of 77–78% and normalisation factors compatible with one considering the quadratic sum of the statistical and experimental systematic uncertainties. The potential DM signal contamination is below 1%.

Top Quark Pair Production in Association with a Vector Boson

As in the search for top squarks, the main contribution in $t\bar{t} + V$ background is arising from $t\bar{t} + Z$ where the Z boson decays into a pair of neutralinos. For estimating the normalisation of $t\bar{t} + Z$ production, the same method as described in Sect. 11.4 is used applying an additional requirement of $m_T\left(\ell, E_T^{miss}\right) > 30$ GeV. The event yields after applying all $t\bar{t} + \gamma$ CR requirements on both data and simulated SM backgrounds as well as the purity and the naive normalisation computed by using Eq. (11.5) are summarised in Table 12.7. The $t\bar{t} + \gamma$ contribution in the CR has a purity of 81% and a normalisation factor compatible with one considering the quadratic sum of the statistical and experimental systematic uncertainties.

Table 12.8 Expected yields of the multi-jet backgrounds in the DM SRs estimated using the jet smearing technique

Region	Expected multi-jet yield
SRt1	0.39 ± 0.15
SRt2	0.15 ± 0.05

Estimation of Multi-jet Contributions

In order to estimate the contributions arising from multi-jet production in the SRs, the jet smearing method introduced in Sect. 11.4 is applied. The expected contribution associated with the multi-jet background in SRt1 and SRt2 is shown in Table 12.8. A requirement on $\left| \Delta\phi \left(\mathrm{jet}^{0-3}, E_{\mathrm{T}}^{\mathrm{miss}} \right) \right| > 0.4$ reduces the contribution to a negligible level. The contribution is smaller in SRt2 due to the selection on $E_{\mathrm{T}}^{\mathrm{miss}}/\sqrt{H_{\mathrm{T}}}$ which further reduces contributions originating from jet mismeasurements. More details on the definitions of the CRs used for the jet smearing method can be found in Appendix C.4.

12.5 Systematic Uncertainties

The estimation of the experimental systematic uncertainties is done with the same procedure as described in Sect. 11.5 resulting in the same dominating experimental uncertainties due to the similarities in the definitions of SRs and CRs.

   For the estimation of the theoretical uncertainties, the same procedure as in the search for top squarks is used for estimating the uncertainties on SM $t\bar{t}$, $t\bar{t} + V$, Z+jets and W+jets production. For the remaining background contributions arising from the production of single top quarks and diboson production, an uncertainty of 50% is assumed.

   In order to estimate the theoretical uncertainties on the signal simulations, the factorisation and renormalisation scales as well as the merging scale of the matrix element and the parton shower are varied by a factor of two and one half with respect to the nominal value. Furthermore, the uncertainty on the choice of the PDF set is estimated by comparing all PDF set included in the NNPDF3.0NNLO description. The quadratic sum of all uncertainties is found to be less than 10%. Thus, for all signal scenarios, a theoretical uncertainty of 10% is assumed.

12.6 Statistical Interpretation

Analogously to the search for top squarks (cf. Sect. 11.6) only the CRs are used to constrain the fit parameters and potential signal contaminations are neglected. The fit results are shown in Table 12.9. The observed data are found to be compatible with the SM predictions. Figure 12.15 shows a comparison between the SM predictions

Table 12.9 Fit results in SRt1 and SRt2 for an integrated luminosity of 36.1 fb^{-1}. The background normalisation parameters are obtained from the background-only fit in the CRs and are applied to the SRs. Small backgrounds are indicated as Others. Benchmark signal models yields ($g = g_t = g_\chi = 1$) are given for each SR. The uncertainties in the yields include statistical uncertainties and all systematic uncertainties defined in Sect. 12.5

	SRt1	SRt2
Observed	23	24
Total background	20.5 ± 5.8	20.4 ± 2.9
$t\bar{t}$	7.0 ± 3.9	3.1 ± 1.3
$t\bar{t}+Z$	4.3 ± 1.1	6.9 ± 1.4
W+jets	3.3 ± 2.6	1.28 ± 0.50
Z/γ^*+ jets	3.7 ± 1.4	6.2 ± 1.1
Others	2.2 ± 1.2	3.00 ± 1.6
Signal scenarios		
$(m_\phi, m_\chi) = (20, 1)$ GeV	9.33 ± 1.63	12.82 ± 1.90
$(m_a, m_\chi) = (20, 1)$ GeV	7.62 ± 1.49	12.11 ± 1.81
$(m_\phi, m_\chi) = (300, 1)$ GeV	2.76 ± 1.15	4.66 ± 1.18
$(m_a, m_\chi) = (300, 1)$ GeV	3.06 ± 1.34	6.20 ± 1.85

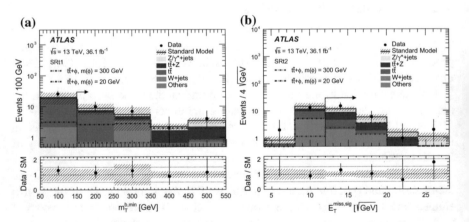

Fig. 12.15 Comparison of the data with the post-fit SM prediction of the $m_T^{b,min}$ distribution in SRt1 (**a**) and the $E_T^{miss}/\sqrt{H_T}$ distribution in SRt2 (**b**). The last bins include overflows, where applicable. All signal region requirements except the one on the distribution shown are applied. The signal region requirement on the distribution shown is indicated by an arrow. The bottom panel shows the ratio of the data to the prediction. The uncertainty band includes all systematic and statistical uncertainties [8]

and the observed data for the distributions of $m_T^{b,min}$ and $E_T^{miss}/\sqrt{H_T}$ in SRt1 and SRt2, respectively, without applying the selection requirement on the corresponding variable. Since no excess above the SM expectation is found, exclusion limits are derived using the CL$_s$ method at 95% CL.

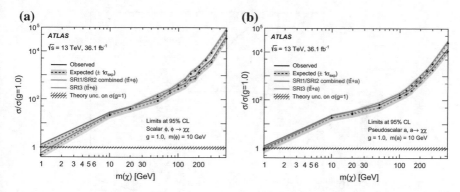

Fig. 12.16 Exclusion limits for $t\bar{t} + \phi$ (**a**) and $t\bar{t} + a$ (**b**) models as a function of the WIMP mass for a mediator mass of 10 GeV. The limits are calculated at 95% CL and are expressed in terms of the ratio of the excluded cross section to the nominal cross section for a coupling assumption of $g = g_t = g_\chi = 1$. The solid (dashed) lines show the observed (expected) exclusion limits for the different SRs according to the colour code specified in the legend. To derive the results for the fully hadronic $t\bar{t}$ final state the region SRt1 or SRt2 providing the better expected sensitivity is used [8]

The exclusion limits on the signal cross section are shown as a function of the WIMP mass assuming the mediator mass to be $m_{\phi,a} = 10$ GeV for both a scalar (cf. Fig. 12.16a) and a pseudo-scalar mediator (cf. Fig. 12.16b) scaled to the signal cross section for a coupling of $g = g_t = g_\chi = 1$. For a scalar mediator and a WIMP mass of $m_\chi = 50$ GeV, 100 times the production cross section for a coupling of $g = 1$ can be excluded (cf. Fig. 12.16a).

For each WIMP and mediator mass pair, the exclusion limit on the production cross section of a scalar mediator can be converted into a limit on the spin-independent DM-nucleon scattering cross section [9]. However, it was found that a SR which is not part of this thesis targeting the dileptonic $t\bar{t}$ decay of the signal signature results in more stringent limits for the conversion into the spin-independent DM-nucleon scattering cross section than exploiting the limits from SRt1 and SRt2 [8]. These limits are labelled as SRt3 in Fig. 12.16.

Figure 12.17 shows the limits on the spin-independent DM-nucleon scattering cross section at 90% CL derived from the exclusion limits on the scalar mediator mass depicted in Fig. 12.16a. The maximum DM-nucleon scattering cross section corresponds to a mediator mass of 10 GeV. For pseudoscalar mediators, the limit on the spin-independent DM-nucleon scattering cross section is several orders of magnitude worse since the DM-nucleon scattering cross sections are suppressed by velocity dependent terms [8]. For WIMP masses below 5 GeV, the results significantly improve the limits on the scattering cross section obtained by the direct-detection experiments (cf. Fig. 12.17). However, the limits from this search are only valid for the theoretical models considered and therefore model-dependent.

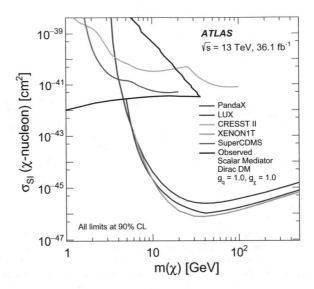

Fig. 12.17 Comparison of the 90% CL limits on the spin-independent DM-nucleon cross-section as a function of DM mass between these results and the direct-detection experiments, in the context of the colour-neutral simplified model with scalar mediator. The black line indicates the exclusion contour derived from the observed limits of SRt3. Values inside the contour are excluded [8]. The exclusion limit is compared with limits from the LUX [10], PandaX-II [11], XENON [12], SuperCDMS [13] and CRESST-II [14] experiments

References

1. Giorgio B et al (2014) On the validity of the effective field theory for dark matter searches at the LHC. Phys Lett B 728:412–421. https://doi.org/10.1016/j.physletb.2013.11.069
2. Busoni G et al (2014) On the validity of the effective field theory for dark matter searches at the LHC, Part II: complete analysis for the *s*-channel. JCAP 1406:060. https://doi.org/10.1088/1475-7516/2014/06/060
3. Busoni G et al (2014) On the validity of the effective field theory for dark matter searches at the LHC Part III: analysis for the *t*-channel. JCAP 1409:022. https://doi.org/10.1088/1475-7516/2014/09/022
4. Abercrombie D et al (2015) Dark matter benchmark models for early LHC Run-2 searches: report of the ATLAS/CMS dark matter forum. In: Boveia A et al (ed). arXiv: 1507.00966 [hep-ex]
5. D'Ambrosio G et al (2002) Minimal flavor violation: an effective field theory approach. In: Nucl Phys B645 155-187. https://doi.org/10.1016/S0550-3213(02)00836-2. arXiv: hep-ph/0207036 [hep-ph]
6. Haisch U, Re E (2015) Simplified dark matter top-quark interactions at the LHC. JHEP 06:078. https://doi.org/10.1007/JHEP06(2015)078
7. Buckley M.R., Feld D, Goncalves D (2015) Scalar simplified models for dark matter. Phys Rev D91:015017. https://doi.org/10.1103/PhysRevD.91.015017
8. ATLAS Collaboration (2018) Search for dark matter produced in association with bottom or top quarks in $\sqrt{s} = 13\,TeV\,pp$ collisions with the ATLAS detector. Eur Phys J C 78:18
9. Busoni G et al (2016) Recommendations on presenting LHC searches for missing transverse energy signals using simplified s-channel models of dark matter. In: Boveia A et al (ed) arXiv: 1603 04156 [hep-ex]

10. Akerib DS et al (2017) Results from a search for dark matter in the complete LUX exposure. Phys Rev Lett 118.2:021303. https://doi.org/10.1103/PhysRevLett.118.021303.
11. Tan A et al (2016) Dark matter results from first 98.7 days of data from the PandaX-II experiment. Phys Rev Lett 117.12:121303. https://doi.org/10.1103/PhysRevLett.117.121303
12. Aprile E et al (2017) First dark matter search results from the XENON1T experiment. Phys Rev Lett 119.18:181301. https://doi.org/10.1103/PhysRevLett.119.181301
13. Agnese R et al (2016) New results from the search for low-mass weakly interacting massive particles with the CDMS Low Ionization Threshold experiment. Phys Rev Lett 116(7 Feb 2016):071301. https://doi.org/10.1103/PhysRevLett.116.071301
14. Angloher G et al (2016) Results on light dark matter particles with a low-threshold CRESST-II detector. Eur Phys J C76.1:25. https://doi.org/10.1140/epjc/s10052-016-3877-3

Chapter 13
The Search for Dark Energy

As introduced in Sect. 3.4, the search for Dark Energy (DE) at the LHC relies on the assumption of a non-zero interaction between the DE and the SM fields. Thereby, evidence for DE production can be found either in precision measurements sensitive to the production of virtual DE particles in loop processes or in the direct production of DE particles in the pp collisions. While the former yields to rather weak constraints [1, 2], the latter is a more effective way for constraining DE models, in especially the ones where DE is produced involving large momentum transfers [3].

An Effective Field Theory (EFT) model provides the most general framework for describing DE theories without knowing the details of the microscopic DE interactions. Following the Horndeski theories [4], the introduction of one scalar field φ with second order equations of motion contains many well-known specific DE models such as quintessence, Galileons or higher dimensional models [5]. The model contains nine operators which are invariant under a shift symmetry $\varphi \rightarrow \varphi + \text{const}$, each suppressed by powers of the characteristic energy scale M of the EFT according to the operator's dimensionality d. The Lagrangian of the EFT can be written as

$$\mathcal{L} = \mathcal{L}_{\text{SM}} + \sum_{i=1}^{9} c_i \mathcal{L}_i = \mathcal{L}_{\text{SM}} + \sum_{i=1}^{9} c_i \frac{\mathcal{O}_i^{(d)}}{M_i^{(d-4)}} \quad , \tag{13.1}$$

where $\mathcal{L}_{\text{SM}}$ is the Lagrangian describing the SM and c_i are so-called Wilson coefficients [5]. The two least suppressed operators are

$$\mathcal{L}_1 = \frac{\partial_\mu \varphi \partial^\mu \varphi}{M^4} T_\nu^\nu \tag{13.2}$$

$$\mathcal{L}_2 = \frac{\partial_\mu \varphi \partial_\nu \varphi}{M^4} T^{\mu\nu} \quad , \tag{13.3}$$

© Springer Nature Switzerland AG 2019
N. M. Köhler, *Searches for the Supersymmetric Partner of the Top Quark, Dark Matter and Dark Energy at the ATLAS Experiment*, Springer Theses, https://doi.org/10.1007/978-3-030-25988-4_13

Fig. 13.1 Feynman
diagrams corresponding to
top quark pair production in
association with a pair of
Dark Energy scalars φ
produced via the $\mathcal{L}_1$ (a) and
$\mathcal{L}_2$ (b) operator, respectively

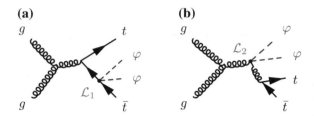

where $T^{\mu\nu}$ is the energy-momentum tensor of the SM Lagrangian. Since $T_\nu^\nu = m\bar{\psi}\psi$
for a Dirac field ψ, the $\mathcal{L}_1$ operator corresponds to a coupling of DE to SM fermions
which is proportional to their Yukawa couplings, while the $\mathcal{L}_2$ operator corresponds to
the kinetic terms of the interaction between DE and SM fermions. The more strongly
suppressed operators $\mathcal{L}_3$–$\mathcal{L}_9$ which are corresponding to higher order versions of $\mathcal{L}_1$
and $\mathcal{L}_2$ are not considered in the scope of this thesis.

Since the coupling of DE to the SM fermions is proportional to their masses [5],
the interaction of DE with top quarks will provide the dominant contribution in case
DE is produced in a pp collision. The Feynman graphs involving the $\mathcal{L}_1$ and $\mathcal{L}_2$
operators are shown in Fig. 13.1.

13.1   Simulation of Dark Energy Production

For the simulation of DE production involving the $\mathcal{L}_1$ and $\mathcal{L}_2$ operators described in
Eq. (13.2), the mass of the DE scalar is set to $m_\varphi = 0.1\,\text{GeV}$ since very small masses
$m_\varphi = \mathcal{O}(H_0) \approx 10^{-42}\,\text{GeV}$ are needed in order to reproduce the correct equation
of state for DE [6]. It was found that at LHC energies, for masses $m_\varphi < 0.1\,\text{GeV}$,
the production cross section is almost independent of m_φ and the kinematics of
the DE field has a negligible effect on the event topology, especially on kinematic
quantities such as the missing transverse energy. Thus, events were generated only
using $M = 400\,\text{GeV}$ ($M = 600\,\text{GeV}$) for setting $c_1 = 1$ ($c_2 = 1$) and all other $c_i = 0$,
since the normalisation of the simulated signal contributions can be rescaled by the
corresponding cross sections for using other characteristic energy scales M.

The signal events are simulated using the MADGRAPH5_aMC@NLO 2.5.5 generator. The parton showering is performed using PYTHIA 8.212 with the A14 set
of underlying-event tuned parameters and the NNPDF2.3LO PDF set. The matrix
elements are calculated at LO in QCD vetoing electroweak contributions and using
the NNPDF3.0LO PDF set. Requiring only one insertion of the $\mathcal{L}_1$ or $\mathcal{L}_2$ operator
into each diagram ensures that the amplitude scales as M^{-4} and thus, only affects the
cross section but no the shape of the kinematic distributions. The simulated events
are processed through the ATLAS detector simulation fully based on GEANT 4.

13.2 Choice of Selection Criteria

The relative number of simulated jets and b-jets after applying a lepton veto is shown in Fig. 13.2 for DE signal events with setting Wilson coefficients to $c_1 = 1$ and $c_2 = 1$, respectively. More than 95% of the simulated DE events contain at least four jets, more than 90% of the events contain at least two b-jets. Looking at the simulated distribution of transverse momentum of the second jet leading in p_T (cf. Fig. 13.3a), more than 92% of the DE events are satisfying $p_T^2 > 80$ GeV while a requirement of $p_T^4 > 40$ GeV rejects almost 20% of the DE signals but also 44% of the SM contributions (cf. Fig. 13.3b).

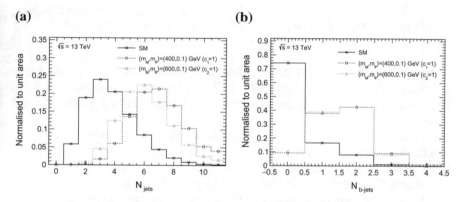

Fig. 13.2 Relative number of simulated jets and b-jets after applying a lepton veto for a selection of dark energy pair production processes and the expected SM contributions

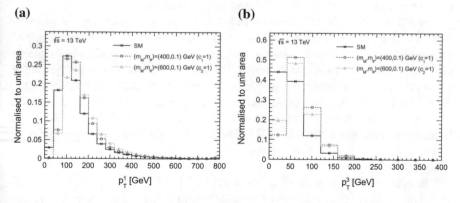

Fig. 13.3 Relative distributions of the simulated jet transverse momenta p_T for the second (**a**) and forth (**b**) jet leading in p_T after applying a lepton veto as well as at least four jets and one b-jet for the same selection of signal processes shown in Fig. 13.2. The distributions of transverse momenta of the other two jets, can be found in Appendix 19.1

(a)   **(b)**

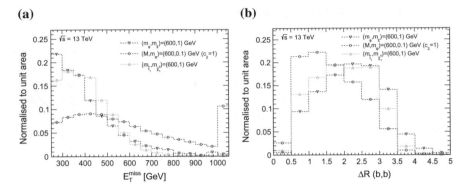

Fig. 13.4 Relative distributions of E_T^{miss} (**a**) and $\Delta R(b, b)$ (**b**) for one simulated Dark Matter (blue), Dark Energy (red) and SUSY (green) scenario. The rightmost bin includes overflow events

As in the search for Dark Matter, the striking similarities in the basic requirements compared to the search for top squarks (cf. Chap. 11) result in the circumstance that the very same basic selection summarised in Table 11.1 is also exploited for this search.

Comparing the shape of the distributions of discriminating variables used in the searches for top squark pair production and Dark Matter, such as E_T^{miss} (cf. Fig. 13.4a) or $\Delta R(b, b)$ (cf. Fig. 13.4b), the decay products of the top quarks are more boosted in case of DE production. This results in a larger E_T^{miss} as well as in rather collimated b-jets compared to top squark or WIMP pair production. In order to estimate the sensitivity for DE production in the final state with jets and missing transverse momentum, the expected DE yields as well as the corresponding expected significances are shown in Tables 13.1, 13.2 and 13.3, respectively.

The SRs in which the DE scenarios obtain the highest expected significances are SRA and SRE of the search for top squark pair production. Since SRA was optimised for large top squark and small neutralino masses, while SRE was optimised for gluino pair production with subsequent decays into top quarks and their supersymmetric partners where the latter are not identified in the detector and thus, contributing to E_T^{miss}. Both scenarios result in strongly boosted jets and large missing transverse momentum which explains the high expected significance of the DE signal scenarios in these regions. Since SRA-TT, SRA-TW and SRA-T0 are disjoint, a statistical combination of them is performed resulting in a higher expected significance than using SRE.

The estimation of the normalisation of the SM contributions is performed identically with respect to the search for top squark pair production (cf. Sect. 11.4) since the SR and CR definitions remain unchanged.

Table 13.1 Expected number of simulated events in SRA and SRB of the search for top squark pair production (cf. Sect. 11.3) scaled to an integrated luminosity of 36.1 fb^{-1}. For the signal scenarios, the expected significance is also given. For the SM processes, the statistical and the experimental systematic uncertainties are shown

	SRA-TT	SRA-TW	SRA-T0	SRB-TT	SRB-TW	SRB-T0
$(M, m_\varphi) = (400, 0.1)$ GeV $(c_1 = 1)$	$0.82 \pm 0.02\ (0.0\sigma)$	$0.76 \pm 0.02\ (0.0\sigma)$	$0.89 \pm 0.02\ (0.0\sigma)$	$1.17 \pm 0.02\ (0.0\sigma)$	$1.34 \pm 0.02\ (0.0\sigma)$	$2.07 \pm 0.03\ (0.0\sigma)$
$(M, m_\varphi) = (600, 0.1)$ GeV $(c_2 = 1)$	$9.56 \pm 0.16\ (2.1\sigma)$	$9.31 \pm 0.16\ (1.9\sigma)$	$23.81 \pm 0.25\ (2.6\sigma)$	$11.39 \pm 0.17\ (0.8\sigma)$	$10.68 \pm 0.17\ (0.5\sigma)$	$22.27 \pm 0.24\ (0.2\sigma)$
$t\bar{t}$	$0.60 \pm 0.10^{+0.174}_{-0.150}$	$0.45 \pm 0.12^{+0.120}_{-0.181}$	$1.45 \pm 0.31^{+0.470}_{-0.452}$	$6.10 \pm 0.59^{+1.107}_{-1.285}$	$12.81 \pm 1.03^{+2.516}_{-3.040}$	$47.23 \pm 2.04^{+8.558}_{-6.928}$
Z + jets	$2.15 \pm 0.27^{+0.640}_{-0.663}$	$4.20 \pm 0.50^{+0.826}_{-0.812}$	$8.63 \pm 0.65^{+1.455}_{-1.154}$	$7.72 \pm 0.66^{+1.234}_{-1.280}$	$14.41 \pm 1.07^{+3.227}_{-2.890}$	$53.99 \pm 1.81^{+8.685}_{-6.960}$
W + jets	$0.65 \pm 0.16^{+0.116}_{-0.082}$	$0.70 \pm 0.20^{+0.147}_{-0.432}$	$1.58 \pm 0.47^{+0.636}_{-0.679}$	$6.12 \pm 2.74^{+3.000}_{-3.081}$	$3.83 \pm 0.61^{+1.093}_{-0.970}$	$20.39 \pm 3.15^{+6.414}_{-6.959}$
Single top	$1.03 \pm 0.55^{+0.421}_{-0.318}$	$0.60 \pm 0.15^{+0.176}_{-0.177}$	$2.52 \pm 1.02^{+1.258}_{-1.176}$	$3.59 \pm 0.64^{+0.618}_{-0.607}$	$5.16 \pm 0.46^{+0.803}_{-0.911}$	$22.53 \pm 1.60^{+2.079}_{-3.372}$
$t\bar{t} + V$	$2.46 \pm 0.28^{+0.277}_{-0.345}$	$1.43 \pm 0.21^{+0.282}_{-0.188}$	$2.02 \pm 0.21^{+0.222}_{-0.227}$	$7.25 \pm 0.50^{+0.902}_{-0.786}$	$8.37 \pm 0.51^{+1.071}_{-0.882}$	$15.92 \pm 0.62^{+1.841}_{-1.346}$
Diboson	$0.12 \pm 0.08^{+0.148}_{-0.129}$	$0.34 \pm 0.14^{+0.072}_{-0.054}$	$0.84 \pm 0.31^{+0.271}_{-0.270}$	$0.23 \pm 0.11^{+0.185}_{-0.149}$	$1.84 \pm 0.57 \pm 0.467$	$2.94 \pm 0.46^{+0.437}_{-0.491}$
Total SM	$7.01 \pm 0.70^{+1.045}_{-1.018}$	$7.72 \pm 0.62^{+1.186}_{-1.118}$	$17.05 \pm 1.38^{+3.521}_{-2.246}$	$31.01 \pm 2.99^{+4.500}_{-4.795}$	$46.43 \pm 1.84^{+8.046}_{-7.823}$	$163.00 \pm 4.53^{+24.198}_{-20.590}$

Table 13.2 Expected number of simulated events in SRD and SRE of the search for top squark pair production (cf. Sect. 11.3) scaled to an integrated luminosity of 36.1 fb^{-1}. The DE signal scenarios do not contribute to the compressed signal regions SRC and thus, SRC is not shown here. For the signal scenarios, the expected significance is also given. For the SM processes, the statistical and the experimental systematic uncertainties are shown

	SRD-low	SRD-high	SRE
$(M, m_\varphi) = (400, 0.1)\,\text{GeV}$ $(c_1 = 1)$	$0.62 \pm 0.01^{+0.063}_{-0.052}\,(0.0\sigma)$	$0.33 \pm 0.01^{+0.042}_{-0.034}\,(0.0\sigma)$	$0.49 \pm 0.01^{+0.046}_{-0.050}\,(0.0\sigma)$
$(M, m_\varphi) = (600, 0.1)\,\text{GeV}$ $(c_2 = 1)$	$6.79 \pm 0.13^{+0.761}_{-0.599}\,(0.6\sigma)$	$4.63 \pm 0.11^{+0.514}_{-0.427}\,(0.9\sigma)$	$10.33 \pm 0.17^{+0.942}_{-0.940}\,(3.3\sigma)$
$t\bar{t}$	$3.43 \pm 0.37^{+0.609}_{-0.695}$	$1.04 \pm 0.20^{+0.230}_{-0.220}$	$0.21 \pm 0.06^{+0.075}_{-0.069}$
$Z + \text{jets}$	$6.68 \pm 0.44^{+1.126}_{-1.051}$	$3.10 \pm 0.27^{+0.580}_{-0.658}$	$1.15 \pm 0.18^{+0.227}_{-0.210}$
$W + \text{jets}$	$4.78 \pm 2.68^{+2.751}_{-2.881}$	$0.84 \pm 0.16^{+0.309}_{-0.348}$	$0.42 \pm 0.13^{+0.159}_{-0.270}$
Single top	$3.30 \pm 0.47^{+0.773}_{-0.831}$	$1.30 \pm 0.22^{+0.183}_{-0.182}$	$0.56 \pm 0.14^{+0.072}_{-0.066}$
$t\bar{t} + V$	$3.06 \pm 0.31^{+0.432}_{-0.411}$	$1.06 \pm 0.15^{+0.222}_{-0.214}$	$0.69 \pm 0.13^{+0.137}_{-0.145}$
Diboson	$0.09 \pm 0.05^{+0.022}_{-0.166}$	$0.48 \pm 0.42^{+0.856}_{-0.728}$	$0.14 \pm 0.08 \pm 0.041$
Total SM	$21.35 \pm 2.80^{+3.785}_{-4.008}$	$7.83 \pm 0.63^{+1.634}_{-1.691}$	$3.17 \pm 0.31^{+0.411}_{-0.531}$

Table 13.3 Expected number of simulated events in the signal regions of the search for WIMP pair production (cf. Sect. 12.3) scaled to an integrated luminosity of 36.1 fb^{-1}. For the DE signal scenarios, the expected significance is also given. For the SM processes, the statistical and the experimental systematic uncertainties are shown

	SRt1	SRt2
$(M, m_\varphi) = (400, 0.1)\,\text{GeV}$ $(c_1 = 1)$	$0.79 \pm 0.02^{+0.069}_{-0.058}\,(0.0\sigma)$	$1.41 \pm 0.02^{+0.133}_{-0.101}\,(0.0\sigma)$
$(M, m_\varphi) = (600, 0.1)\,\text{GeV}$ $(c_2 = 1)$	$9.16 \pm 0.15^{+0.817}_{-0.813}\,(1.0\sigma)$	$12.42 \pm 0.18^{+1.076}_{-1.071}\,(1.4\sigma)$
$t\bar{t}$	$6.12 \pm 0.50^{+1.376}_{-1.105}$	$2.79 \pm 0.30^{+0.559}_{-0.630}$
$Z + \text{jets}$	$3.24 \pm 0.34^{+0.717}_{-0.991}$	$5.72 \pm 0.72^{+1.226}_{-1.182}$
$W + \text{jets}$	$3.31 \pm 2.67^{+2.990}_{-3.071}$	$1.28 \pm 0.45^{+0.291}_{-0.218}$
Single top	$1.33 \pm 0.21^{+0.276}_{-0.266}$	$1.99 \pm 0.30^{+0.307}_{-0.645}$
$t\bar{t} + V$	$3.53 \pm 0.34^{+0.382}_{-0.345}$	$5.60 \pm 0.44^{+0.809}_{-0.579}$
Diboson	$0.45 \pm 0.15^{+0.196}_{-0.224}$	$0.86 \pm 0.24^{+0.214}_{-0.216}$
Total SM	$17.97 \pm 2.77^{+3.246}_{-3.399}$	$18.24 \pm 1.07^{+2.698}_{-2.631}$

13.3 Systematic Uncertainties

Since the search for DE is exploiting SRA-TT, SRA-TW and SRA-T0 from the search for top squark pair production, the SM contributions and their uncertainties can be directly taken from the estimates performed in Sect. 11.5. In order to estimate the theoretical uncertainties on the simulation of DE production, one distinguishes between uncertainties affecting the normalisation of the signal simulation and uncertainties depending on the selection criteria. Uncertainties arise from the choice of the PDF set, the renormalisation and factorisation scale and the strong coupling constant α_s. The uncertainties on those choices are estimated by varying the nominal values of the parameters by a factor of two up and down, and retrieving the event yields in order to be able to calculate the transfer factors introduced in Sect. 11.5. The quadratic sum of the uncertainties depending on the selection criteria is 12% (9%) for the model involving the $\mathcal{L}_1$ ($\mathcal{L}_2$) operator, respectively. Those values are used in the profile likelihood fit.

The theoretical uncertainties on the signal simulation affecting the normalisation only are propagated to the uncertainty band of the cross section limits (cf. Sect. 13.4).

13.4 Statistical Interpretation

A simultaneous fit of SRA-TT, SRA-TW and SRA-T0 is performed in order to obtain an upper limit on the production cross section of DE. The generated DE signals with $M = 400 \, \text{GeV}$ ($M = 600 \, \text{GeV}$) for the $\mathcal{L}_1$ ($\mathcal{L}_2$) operator are rescaled to other suppression scales M since the amplitude of the matrix element scales as M^{-4} and thus, M just affects the production cross section but not the event topology. From the upper limits on the production cross sections for different suppression scales M, a lower limit on M can be derived depending on the production cross section. Figure 13.5 shows the lower limits on the suppression scale M of the operators $\mathcal{L}_1$ and $\mathcal{L}_2$ obtained by combining SRA-TT, SRA-TW and SRA-T0. For DE production involving the $\mathcal{L}_1$ ($\mathcal{L}_2$) operator, suppression scales up to $M < 309 \, \text{GeV}$ ($M < 674 \, \text{GeV}$) can be excluded at 95% CL.

13.5 Validity of the EFT Approximation

In order to ensure the validity of an EFT with a suppression scale of M, the momentum transfer present in the interaction, Q_{tr}, has to be well below the suppression scale, $Q_{\text{tr}} \ll M$. In general, an EFT approximation should be valid as long as

$$Q_{\text{tr}} < g_* \cdot M \quad , \tag{13.4}$$

(a)   **(b)**

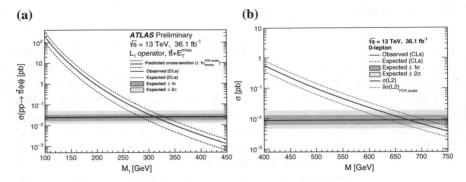

Fig. 13.5 Lower limit on the suppression scale M of the operators $\mathcal{L}_1$ (**a**) and $\mathcal{L}_2$ (**b**) at 95% CL obtained by combining SRA-TT, SRA-TW and SRA-T0 and rescaling the simulated DE scenarios to different suppression scales [7]. The theoretical uncertainties on the choice of the PDF set and the renormalisation and factorisation scales of the signal simulation affecting the normalisation are drawn as the dashed uncertainty bands

where $g_* < 4\pi$, for the couplings to be in a perturbative regime [8]. The momentum transfer can be calculated by [9]

$$Q_{\text{tr}} = \sqrt{\hat{s}} = \sum_{i \in \text{interacting partons}} p_i \quad . \tag{13.5}$$

In case events do not satisfy Eq. 13.4, the calculated cross section limits have to be rescaled in order to obtain valid results [10]. For estimating the rescaling factor, the fraction of events satisfying Eq. 13.4 is defined as

$$R_{\text{valid}} = \frac{N(Q_{\text{tr}} < Q_{\text{tr}}^{\max} = g_* M)}{N} \tag{13.6}$$

where $Q_{\text{tr}}^{\max}$ is the maximum momentum transfer measured in all events. The fraction evaluated for all simulated events is referred to as $R_{\text{valid}}^{\text{tot}}$. The following iterative calculation is performed to obtain the rescaled suppression scale M_{rescaled}:

- Assume that all events are valid in terms of the EFT, especially, Eq. (13.6) reads $R_{\text{valid},i=1} = 1$ in the first iteration.
- Calculate $Q_{\text{tr},i}^{\max} = g_* M_i = g_* M_{i-1}$.
- Calculate $R_{\text{valid},i} = \frac{N(Q_{\text{tr}} < Q_{\text{tr},i}^{\max})}{N(Q_{\text{tr}} < Q_{\text{tr},i-1}^{\max})}$.
- Calculate $M_i = R_{\text{valid}}^{\text{tot}} \cdot M_i$.
- Continue the iteration until $R_{\text{valid},i} = 0$ or $R_{\text{valid},i} = 1$.

The final rescaled suppression scale is obtained by $M_{\text{rescaled}} = \prod_i R_{\text{valid},i} \cdot M$.

Figure 13.6 shows the exclusion contours of the rescaled suppression scale M of the operators $\mathcal{L}_1$ and $\mathcal{L}_2$ drawn against the effective coupling associated to the UV completion of the EFT, g_*.

(a)   **(b)**

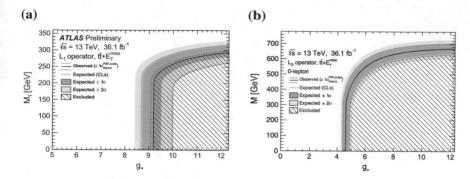

Fig. 13.6 Exclusion contours of the suppression scale M of the operators $\mathcal{L}_1$ (**a**) and $\mathcal{L}_2$ (**b**) drawn against the effective coupling associated to the UV completion of the EFT, g_* [7]

For DE production involving the $\mathcal{L}_2$ operator (cf. Fig. 13.6b), suppression scales $M < 670\,\text{GeV}$ can be excluded for effective couplings $g_* \geq 3.3\pi$. However, more stringent limits on the suppression scale for the $\mathcal{L}_2$ operator can be derived in a selection targeting the presence of one single highly energetic jet [7] which will not be discussed here. The limits on the suppression scale of the $\mathcal{L}_1$ operator (cf. Fig. 13.6b) are currently the most stringent ones but the search is not yet sensitive to weakly coupled models.

References

1. Brax P et al (2009) Collider constraints on interactions of dark energy with the standard model. JHEP 09, 128. https://doi.org/10.1088/1126-6708/2009/09/128
2. Brax P et al (2010) Higgs production as a probe of chameleon dark energy. Phys Rev D81:103524. https://doi.org/10.1103/PhysRevD.81.103524
3. Brax P, Burrage C, Englert C (2015) Disformal dark energy at colliders. Phys Rev D 92(4):044036 (2015). https://doi.org/10.1103/PhysRevD.92.044036
4. Horndeski GW (1974) Second-order scalar-tensor field equations in a four-dimensional space. Int J Theor Phys 10:363–384. https://doi.org/10.1007/BF01807638
5. Brax P et al (2016) LHC signatures of scalar dark energy. Phys Rev D 94(8):084054 (2016). https://doi.org/10.1103/PhysRevD.94.084054
6. Brax P (2018) What makes the universe accelerate? a review on what dark energy could be and how to test it. Rep Progr Phys 81(1):016902 (2018)
7. ATLAS Collaboration (2019) Constraints on mediator-based dark matter and scalar dark energy models using ps = 13 TeV pp collision data collected by the ATLAS detector. JHEP 05, 142 (2019). https://doi.org/10.1007/JHEP05
8. Busoni Giorgio et al (2014) On the validity of the effective field theory for dark matter searches at the LHC. Phys Lett B 728:412–421. https://doi.org/10.1016/j.physletb.2013.11.069
9. Englert Christoph, Spannowsky Michael (2015) Effective theories and measurements at colliders. Phys Lett B 740:8–15. https://doi.org/10.1016/j.physletb.2014.11.035
10. Abercrombie D et al (2015) Dark matter benchmark models for early LHC Run-2 searches: report of the ATLAS/CMS dark matter forum. In: Boveia A et al (ed) arXiv:1507.00966 [hep-ex]

Chapter 14
Sensitivity Studies Exploiting Multivariate Techniques

The classification between signal-like and background like-events exploiting discriminating variables as described in Chap. 10 is considered as the most physically intuitive method in searches for new particles or precision measurements in high energy physics. It allows for a good understanding and interpretation of the underlying physics processes as well as the possibility for theoreticians to exploit the event selection for further studies. However, the performance of event classification based on discriminating variables can suffer from correlations amongst a subset of those variables which complicate the definition of the ideal selection requirements in order to thoroughly separate signal-like from background-like events.

Multivariate techniques have become a popular supplement in order to classify events in addition to the commonly used event selections. In particular, the classification algorithms of Boosted Decision Trees [1, 2] and Artificial Neural Networks [3] are of special interest in high energy physics applications [4].

This chapter studies the expected reach in sensitivity for the search of fully-hadronically decaying top squarks (cf. Chap. 11) exploiting multivariate techniques in the context of supervised learning. Supervised learning refers to the fact that the true classification of the input dataset is known (in this case, by using simulated data). Thus, the performance of the multivariate algorithm can be evaluated and optimised. Commonly, this is done by splitting the dataset into a so-called *training* and *testing* subset. This allows for the application of the trained algorithm on the testing dataset and the evaluation of its performance. Unsupervised learning algorithms are not discussed in this thesis. Boosted Decision Trees are introduced in Sect. 14.1 and their application targeting SRs A, B and C are studied. Section 14.2 introduces Artificial Neural Networks, in particular, Multi-Layer Perceptrons, and their potential in signal scenarios with high top squark and low neutralino masses.

© Springer Nature Switzerland AG 2019
N. M. Köhler, *Searches for the Supersymmetric Partner of the Top Quark, Dark Matter and Dark Energy at the ATLAS Experiment*,
Springer Theses, https://doi.org/10.1007/978-3-030-25988-4_14

14.1   Boosted Decision Trees

A decision tree is a binary tree consisting of nodes which classify an event to be signal-like or background-like, respectively, for a fixed set of discriminating variables. The classification at each node is done based on a cut on a discriminating variable where the best choice of both the variable and the corresponding cut value is determined by minimising the so-called Gini index

$$G = w_{\text{pass}} \cdot p_{\text{pass}} \cdot (1 - p_{\text{pass}}) + w_{\text{fail}} \cdot p_{\text{fail}} \cdot (1 - p_{\text{fail}}) \tag{14.1}$$

$$= \frac{N_{\text{pass}}}{N_{\text{pass}} + N_{\text{fail}}} \cdot \frac{S_{\text{pass}} \cdot B_{\text{pass}}}{N_{\text{pass}}^2} + \frac{N_{\text{fail}}}{N_{\text{pass}} + N_{\text{fail}}} \cdot \frac{S_{\text{fail}} \cdot B_{\text{fail}}}{N_{\text{fail}}^2} \tag{14.2}$$

with $w_i = N_i/(N_{\text{pass}} + N_{\text{fail}})$ being the relative fraction of events and $p_i = S_i/(S_{\text{pass}} + B_{\text{fail}})$ the signal purities obtained after passing or failing the corresponding cut ($i \in \{\text{pass, fail}\}$). S_i (B_i) denotes the relative fraction of signal (background events) and $N_{\text{pass}} + N_{\text{fail}}$ the number of events available from the parent node. To avoid decision trees memorising the properties of single events, the depth of the decision tree is limited to a maximum number and the presence of a minimum amount of events at each node is required. This property is commonly referred to as the minimum node size.

However, a single decision tree would not result in any improvement on the signal and background classification since it is nothing else than a usual event selection. Therefore, the concept of boosting [5] is introduced. A misclassification error $\delta_{\text{miscl.}}$ of the decision tree can be defined as the number of misclassified events divided by all events. A weight of

$$w = \exp(\alpha) = \left(\frac{1 - \delta_{\text{miscl.}}}{\delta_{\text{miscl.}}} \right)^{\beta}, \tag{14.3}$$

where β is called the *learning rate* is assigned to all misclassified events of the decision tree and subsequently, all events are reweighted accordingly to preserve the total normalisation. Iteratively creating decision trees by minimising the Gini index and reweighting misclassified events is referred to as the *Adaptive Boost* algorithm [6]. Consequently, the classification algorithm is called Boosted Decision Tree (BDT). The final classification of an event after a successful training phase is achieved by the so-called BDT score defined as [7]

$$\text{BDT score} = \sum_{i=0}^{N_{\text{Trees}}} \frac{\alpha_i \cdot q_i}{\alpha_i}, \tag{14.4}$$

where q_i is indicating whether the event is classified as signal-like or background-like by the i-th tree,

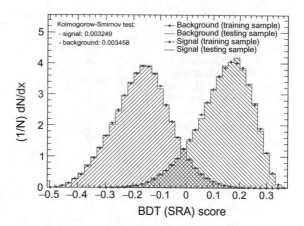

Fig. 14.1 Comparison of the BDT scores obtained for signal and background in the BDT training and testing phase, respectively. In addition, the results of the Kolmogorov–Smirnov test for both signal and background events are depicted. Both trainings are found to be not overtrained

$$q_i = \begin{cases} 1 & \text{if the } i\text{-th tree classifies the event as signal-like.} \\ -1 & \text{if the } i\text{-th tree classifies the event as background-like.} \end{cases} \tag{14.5}$$

Figure 14.1 depicts the distributions of the BDT scores obtained for background-like and signal-like events after the training phase (dotted markers). Evaluating the BDT score for the events of the testing dataset, the hashed areas depicted in Fig. 14.1 which are referred to as *testing* distributions are obtained. In case a large discrepancy is observed between the BDT score of the training and the testing phase, the BDT is declared to be *overtrained*. To avoid overtrain and unnecessary additional computing time, the number trees creating during the boosting is restricted to a maximum number which ensures a saturating misclassification error.

The amount of overtraining can be estimated by performing a two-sample Kolmogorov–Smirnov test [8, 9] defined as

$$k = \max_{x \in [-1,1]} |F_{\text{train}}(x) - F_{\text{test}}(x)| \tag{14.6}$$

with $F_i(x)$ ($i \in \{\text{train, test}\}$) being the cumulative normalised BDT score probability distributions

$$F(x) = \frac{\text{number of events with BDT score } \leq x}{\text{number of all events}} \tag{14.7}$$

which can be calculated for signal and background events, respectively. The BDT is declared to be not overtrained at the level of γ [10] in case of

$$k \cdot \sqrt{\frac{N_{\text{train}} \cdot N_{\text{test}}}{N_{\text{train}} + N_{\text{test}}}} \leq \sqrt{-\frac{1}{2} \cdot \ln\left(\frac{\gamma}{2}\right)}. \tag{14.8}$$

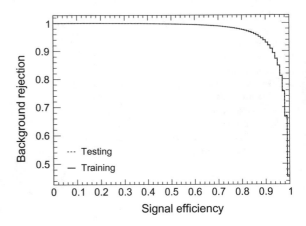

Fig. 14.2 Receiver Operating Characteristic (ROC) curve obtained in the training of a BDT trained on the signal simulations with $900\,\text{GeV} \leq m_{\tilde{t}_1} < 1200\,\text{GeV}$ and $m_{\tilde{\chi}_1^0} < 200\,\text{GeV}$ using all variables of SRA and SRB of the search for top squarks

In the example of Fig. 14.1, the Kolmogorov–Smirnov test is evaluated to be 0.003249 (0.003458) for signal (background) events. With the training and testing datasets comprising 80493 (430185) events each for signal (background) one obtains $\gamma <$ 0.86 ($\gamma < 0.01$). Thus, the BDT is not overtrained at the level of 86% (1%) for the signal (background) events.

Another import characteristic for the evaluation of multivariate classification techniques is the so-called Receiver Operating Characteristic (ROC) curve [11] which shows the background rejection against the signal efficiency (cf. Fig. 14.2). The area under the ROC curve, abbreviated AUC, is a single scalar value representing the expected performance of an event classification algorithm and therefore is a useful quantity for comparing different classifiers [12].

Previous sensitivity studies of the search for top squarks in the fully-hadronic decay channel suggest that a potential gain in sensitivity can be achieved by exploiting BDTs trained on all variables used in the definition of a particular SR [13]. Furthermore, it was found that the usage of several simulated signal scenarios with similar event topologies are significantly improving the sensitivity when exploiting multivariate techniques since the machine learning algorithms require a sufficient amount of statistics in order to outperform the common event selections based on discriminating variables.

High Top Squark Masses

For studying the potential increase in sensitivity by using multivariate techniques for scenarios with high top squark masses, a BDT is trained on all variables used in the selection criteria of SRA and SRB after applying the selection summarised in Table 14.1. The ROOT toolkit for Multivariate Data Analysis (TMVA) [14] is used and the results are validated using SCIKIT- LEARN [15]. Several combinations of signal scenarios are used in the training to estimate the best trade-off between signal statistics needed by the multivariate algorithm and the change of the event kinematics for different signal scenarios. The simulated shape of all distributions of

Table 14.1 Preselection applied for the BDT training targeting SRA, SRB and SRD

	Requirement
Number of leptons	$N_\ell = 0$
Trigger	E_T^{miss}-trigger as described in Table 11.1
Missing transverse energy	$E_T^{miss} > 250\,GeV$
Number of jets	$N_{jets} \geq 4$
Number of b-jets	$N_{b-jets} \geq 1$
$\left\|\Delta\phi\left(jet^{0,1,2}, E_T^{miss}\right)\right\|$	> 0.4
Track-based missing transverse energy $E_T^{miss,track}$	$> 30\,GeV$
$\left\|\Delta\phi\left(E_T^{miss}, E_T^{miss,track}\right)\right\|$	$< \frac{\pi}{3}$
Number of $R = 1, 2$ reclustered jets	$N_{jet, R=1.2}^0 \geq 2$
Number of $R = 0.8$ reclustered jets	$N_{jet, R=0.8} \geq 1$
τ-veto	$N_\tau = 0$

Table 14.2 BDT parameter settings optimised for the search for fully-hadronically decaying top squarks. The optimisation was performed using all discriminating variables of SRA and SRB as well as all simulated samples with $m_{\tilde{t}_1} > 1000\,GeV$ in the BDT training [13]. The number of cuts refers to the granularity of which every variable is scanned when computing the Gini index

Parameter	Setting
Number of trees	1200
Minimum node size	1.5%
Maximum depth of tree	3
Boost type	Adaptive boost
Learning rate β	0.5
Separation type	Gini index
Number of cuts	20

discriminating variables exploited in the BDT training is in agreement with recorded pp collision data (cf. Figs. E.1–E.4).

The chosen parameter settings of the BDT were optimised in previous studies [13] and are summarised in Table 14.2.

After training the BDT, a scan through the BDT score is performed to calculate the expected significances for a potential SR based on the preselection summarised in Table 14.1. The expected significances were calculated assuming an uncertainty for 30% on the expected SM yield and an integrated luminosity of $150.0\,fb^{-1}$ corresponding to the expected LHC Run 2 dataset.

The training model considering all signal scenarios with $900\,GeV \leq m_{\tilde{t}_1} < 1200\,GeV$ and $m_{\tilde{\chi}_1^0} < 200\,GeV$ results in the highest sensitivity for the phase space of high top squark and low neutralino masses (cf. Fig. 14.4a). Furthermore, targeting high top squark and high neutralino masses, the same BDT was trained considering

(a)   **(b)**

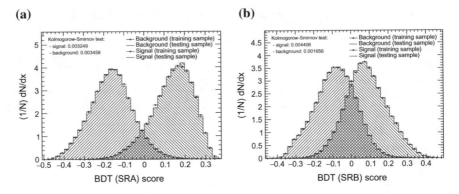

Fig. 14.3 Comparisons of the BDT scores obtained for signal and background in the BDT training and testing phase, respectively, for the BDTs exploited for SRs A (cf. Fig. 14.5b) and B (cf. Fig. 14.6b). In addition, the results of the Kolmogorov–Smirnov test for both signal and background events are depicted. Both trainings are found to be not overtrained

(a)   **(b)**

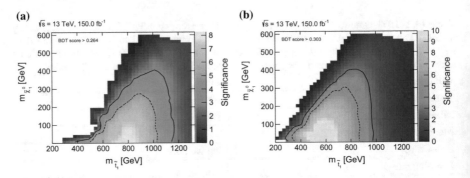

Fig. 14.4 Expected significances as a function of $m_{\tilde{t}_1}$ and $m_{\tilde{\chi}_1^0}$ for signal regions based on the preselection summarised in Table 14.1 and a requirement on the score of a BDT trained on the signal simulations with $900\,\text{GeV} \leq m_{\tilde{t}_1} < 1200\,\text{GeV}$ and $m_{\tilde{\chi}_1^0} < 200\,\text{GeV}$ **(a)** and on all signal simulations available for the $\tilde{t}_1 \rightarrow t + \tilde{\chi}_1^0$ decay scenario **(b)**. The results are obtained assuming an integrated luminosity of $150.0\,\text{fb}^{-1}$ corresponding to the full LHC Run 2 dataset. For the calculation of the expected significances, an uncertainty of 30% on the SM yields are assumed. The solid (dashed) black line indicates the 3σ (5σ) contour

all signal scenario with $\tilde{t}_1 \rightarrow t + \tilde{\chi}_1^0$ decays (cf. Fig. 14.4b). The cut values on the BDT score which give the highest sensitivities are BDT (SRA) > 0.264 (BDT (SRB) > 0.303). Performing the same training exploiting SCIKIT- LEARN gives comparable results validating the usage of TMVA.

To compare the expected gain in sensitivity to the nominal SRA and SRB definitions summarised in Chap. 11, the expected exclusion limits at 95% CL are calculated without exploiting any CRs and assigning a fixed systematic uncertainty of 30%. Figure 14.5a shows the expected exclusion limits at 95% CL for SRA-TT obtained from this simplified exclusion fit. The simplified exclusion fit is validated against the full fit-procedure described in Sect. 11.6 which is used for the 36.1 fb^{-1}

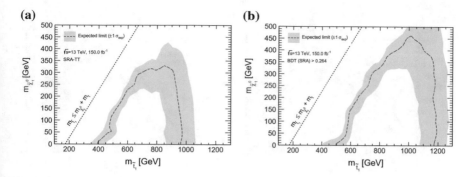

Fig. 14.5 Comparison between the expected exclusion limits at 95% CL for SRA-TT obtained by a simplified maximum likelihood fit assuming an overall systematic uncertainty of 30% and no dedicated CRs (**a**) and a SR based on the preselection summarised in Table 14.1 and a requirement on the score of a BDT trained on the signal simulations with $900\,\text{GeV} \le m_{\tilde{t}_1} < 1200\,\text{GeV}$ and $m_{\tilde{\chi}_1^0} < 200\,\text{GeV}$ (**b**). The results are obtained assuming an integrated luminosity of $150.0\,\text{fb}^{-1}$ corresponding to the full LHC Run 2 dataset

of recorded data (cf. Sect. E.1 in the Appendix). The simplified exclusion fit results in slightly weaker exclusion limits with larger systematic uncertainties which justifies its usage in the sensitivity studies as a conservative approach. The same observation is made for all other SRs (cf. Appendix E). Figure 14.5b shows the corresponding expected exclusion limits with the same setup but requiring a minimum BDT score of BDT (SRA) > 0.264 on top of the preselection summarised in Table 14.1. The application of the SR exploiting the BDT results in an improvement of approximately 200 GeV (20%) towards higher top squark masses for $m_{\tilde{\chi}_1^0} = 1\,\text{GeV}$. However, the application of the BDT-based SR leads to a decrease of sensitivity towards the region $m_{\tilde{t}_1} - m_{\tilde{\chi}_1^0} \sim m_t$.

For high neutralino masses, the expected exclusion limits requiring BDT (SRB) > 0.303 as well as the limits obtained from the statistical combination of SRB-TT, SRB-TW and SRB-T0 are shown in Fig. 14.6. For $m_{\tilde{t}_1} \sim 800\,\text{GeV}$, an expected improvement of approximately 100 GeV (30%) in the neutralino mass can be observed. The good agreement between the BDT scores of the training and testing datasets confirmed by the small value of the Kolmogorov–Smirnov score show that the BDT was not overtrained (cf. Fig. 14.3).

Compressed Scenarios

For signal scenarios with $m_{\tilde{t}_1} - m_{\tilde{\chi}_1^0} \sim m_t$, the nominal SRs SRC1-5 are exploiting the discriminating variables obtained from the recursive jigsaw algorithm. A BDT with the parameter settings summarised in Table 14.2 was trained on all variables used in the selection criteria of SRC1-5 after applying the basic selection summarised in Table 14.3. The distributions of the discriminating variables after applying the preselection are shown in Figs. E.9 and E.10. As for the variables exploited in the BDTs targeting SRA and SRB, the shape of recorded and simulated data are in reasonable agreement.

(a)   **(b)**

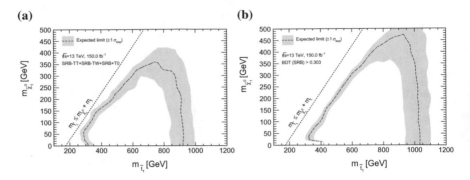

Fig. 14.6 Comparison between the expected exclusion limits at 95% CL for the combination of SRB-TT, SRB-TW and SRB-T0 obtained by a simplified maximum likelihood fit assuming an overall systematic uncertainty of 30% and no dedicated CRs (**a**) and a SR based on the preselection summarised in Table 14.1 and a requirement on the score of a BDT trained on all signal simulations available for the $\tilde{t}_1 \to t + \tilde{\chi}_1^0$ decay scenario (**b**). The results are obtained assuming an integrated luminosity of 150.0 fb^{-1} corresponding to the full LHC Run 2 dataset

Table 14.3 Preselection applied for the BDT training targeting SRC

	Requirement
Number of leptons	$N_\ell = 0$
Trigger	E_T^{miss}-trigger as described in Table 11.1
Missing transverse energy	$E_T^{miss} > 250$ GeV
Number of jets	$N_{jet}^S \geq 5$
Number of b-jets	$p_{T,b}^{0,S} \geq 1$
$\left\|\Delta\phi\left(\text{jet}^{0,1}, E_T^{miss}\right)\right\|$	>0.4
Track-based missing transverse energy $E_T^{miss,track}$	>30 GeV
$\left\|\Delta\phi\left(E_T^{miss}, E_T^{miss,track}\right)\right\|$	$<\frac{\pi}{3}$

The best expected significances are obtained for using all signal simulations with a mass difference between top squark and neutralino in the range 173 GeV $\leq m_{\tilde{t}_1} - m_{\tilde{\chi}_1^0} \leq 188$ GeV in the training of the BDT. To estimate the best cut value on the BDT score, a scan through the BDT score was performed calculating the expected significances for all signal simulations of $\tilde{t}_1 \to t + \tilde{\chi}_1^0$ decays for an integrated luminosity of 150.0 fb^{-1} assuming an uncertainty of 30% on the SM yields. Requring BDT (SRC) > 0.095 was found to result in the highest expected sensitivity for compressed signal scenarios (cf. Fig. 14.7). Looking at the expected exclusion limits at 95% CL assuming an uncertainty of 20%,[1] a SR exploiting a BDT does not result in a significantly improved limit as for SRA and SRB (cf. Fig. 14.8). This can be explained by the fact that the discriminating variables obtained by the recursive

---

[1] For SRC, an uncertainty of 20% is rather corresponding to the uncertainties found in the nominal exclusion fit described in Sect. 11.6 (cf. Fig. E.11 in the Appendix).

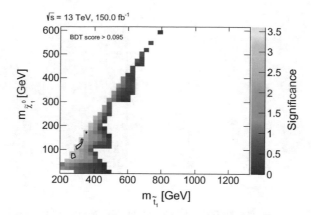

Fig. 14.7 Expected significances as a function of $m_{\tilde{t}_1}$ and $m_{\tilde{\chi}_1^0}$ for a signal region based on the preselection summarised in Table 14.3 and a minimum requirement on the score of a BDT of BDT (SRC) score > 0.095 trained on all signal simulations with $173\,\text{GeV} \leq m_{\tilde{t}_1} - m_{\tilde{\chi}_1^0} \leq 188\,\text{GeV}$. The results are obtained assuming an integrated luminosity of $150.0\,\text{fb}^{-1}$ corresponding to the full LHC Run 2 dataset. For the calculation of the expected significances, an uncertainty of 30% on the SM yields are assumed. The solid black line indicates the 3σ contour

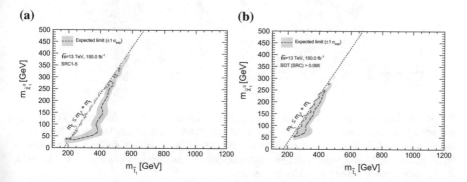

Fig. 14.8 Comparison between the expected exclusion limits at 95% CL for the combination of SRC1-5 obtained by a simplified maximum likelihood fit assuming an overall systematic uncertainty of 20% and no dedicated CRs (**a**) and a SR based on the preselection summarised in Table 14.3 and a requirement on the score of a BDT trained on all signal simulations with $173\,\text{GeV} \leq m_{\tilde{t}_1} - m_{\tilde{\chi}_1^0} \leq 188\,\text{GeV}$ (**b**)

jigsaw algorithm are correlated in the same way for background-like and signal-like events (cf. Fig. 14.9). Therefore, exploiting a multivariate technique does not lead to any improvement. The distributions of the correlated variables are depicted in Fig. E.13 in the Appendix. No visible difference in shape between signal and background is observed. Furthermore, the lower statistics of simulated signal scenarios for $m_{\tilde{t}_1} - m_{\tilde{\chi}_1^0} \sim m_t$, results in a slight overtraining of the BDT targeting SRC (cf. Fig. 14.10) which discourages from exploiting BDTs for the compressed scenario.

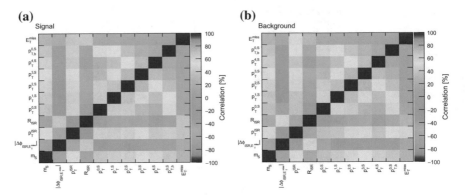

Fig. 14.9 Correlation matrices for signal (**a**) and background (**b**) events for the BDT targeting SRC

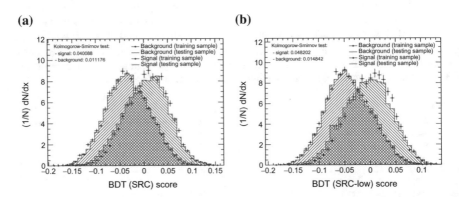

Fig. 14.10 Comparisons of the BDT scores obtained for signal and background in the BDT training and testing phase, respectively, for the BDTs trained on all signal simulations with $173\,\text{GeV} \le m_{\tilde{t}_1} - m_{\tilde{\chi}_1^0} \le 188\,\text{GeV}$ (**a**) as well as only using simulations with $m_{\tilde{t}_1} \le 400\,\text{GeV}$ (**b**). In addition, the results of the Kolmogorov–Smirnov test for both signal and background events are depicted. Both trainings are found to be slightly overtrained

14.2 Artificial Neural Networks

One of the most popular types of multivariate algorithms are Artificial Neural Networks (ANN) [16]. Inspired by the human brain, an ANN consists of several neurons organised in so-called layers and connected amongst each other with analogues of the synapses inside the brain. The strength of the connection to the following neuron is described by weights whose values are optimised during the training phase of the ANN. In case of classification problems, the final layer of the ANN, which is referred to as the output layer, typically consists of one single neuron. Layers in between the first layer and the output layer are commonly referred to as hidden layers. Every ANN with at least one hidden layer can be denoted as *deep neural network* (DNN).

One of the most common algorithms for ANNs is the so-called Multi-Layer Perceptron (MLP) [17, 18] which utilises a supervised learning technique called *backpropagation* for the training [19]. In the MLP, neurons of a given layer can only be connected with neurons from a neighbouring layer. The value which a neuron is propagating is obtained by summing the weighted outputs of previous neurons to the synapse function and connecting the result to the so-called *activation function* to detect potential (non-)linear correlations between the input variables [20]. One of the possible choices for the activation function is the so-called Rectified Linear Units (ReLU) function [21] defined as

$$f(x) = \max\{0, x\}. \tag{14.9}$$

The usage of labelled data allows to define the error function of N_{train} training events by

$$E(\mathbf{v}_1, \ldots, \mathbf{v}_{N_{\text{train}}} | \overleftrightarrow{w}) = \frac{1}{2} \sum_{i=1}^{N_{\text{train}}} \left(y_i - y_i^{\text{true}} \right)^2 , \tag{14.10}$$

where $\mathbf{v}_i$ are the training variables of the i-th event, $\overleftrightarrow{w}$ is the matrix of weights for every event and training variable, y_i is response of the NN to the event and y_i^{true} is the true information whether the event is a signal or a background event

$$y_i^{\text{true}} = \begin{cases} 1 & \text{if the } i\text{-th event is a signal event.} \\ 0 & \text{if the } i\text{-th event is a background event.} \end{cases} \tag{14.11}$$

The ANN ist trained by minimising the error function. Due to its performance, the currently most favoured minimisation algorithm, that can be used instead of the classical stochastic gradient descent procedure to update network weights iteratively based on the training data, is the Adam algorithm [22]. The recommended configuration parameters for the Adam algorithm are the learning rate $\alpha = 0.001$ and the exponential decay rates for the first and seconds moment estimates, $\beta_1 = 0.9$ and $\beta_2 = 0.999$, respectively [22].

Since previous studies revealed that training ANNs on the final discriminating variables of the SRs does not yield convincing improvements in the expected sensitivity as it is the case for BDTs [13], the usage of more basic event properties in ANNs is studied based on SCIKIT- LEARN. Therefore, only the four momenta of the four jets leading in p_T including the b-tagging information and the absolute value and azimuthal angle of the missing transverse momentum vector are chosen as training variables after the preselection shown in Table 14.4 are applied. To optimise the configuration parameters of the MLP based on the ROC AUC, a two-dimensional scan on the number of neurons per layer and the number of training iterations is performed (cf. Fig. 14.11). The configuration with the best agreement between training and testing results is obtained for using two hidden layers with 150 and 10 neurons each and a maximum number of training iterations of 200. However, the values obtained for the ROC AUC are still of similar size which means the performance of

Table 14.4 Preselection applied for the ANN training targeting SRA

	Requirement
Number of leptons	$N_\ell = 0$
Trigger	$E_{\mathrm{T}}^{\mathrm{miss}}$-trigger as described in Table 11.1
Missing transverse energy	$E_{\mathrm{T}}^{\mathrm{miss}} > 250\,\mathrm{GeV}$
Number of jets	$N_{\mathrm{jets}} \geq 4$
Number of b-jets	$N_{b-\mathrm{jets}} \geq 1$
$\left\lvert \Delta\phi\left(\mathrm{jet}^{0,1,2}, E_{\mathrm{T}}^{\mathrm{miss}}\right)\right\rvert$	> 0.4
Track-based missing transverse energy $E_{\mathrm{T}}^{\mathrm{miss,track}}$	$> 30\,\mathrm{GeV}$
$\left\lvert \Delta\phi\left(E_{\mathrm{T}}^{\mathrm{miss}}, E_{\mathrm{T}}^{\mathrm{miss,track}}\right)\right\rvert$	$< \frac{\pi}{3}$
τ-veto	yes

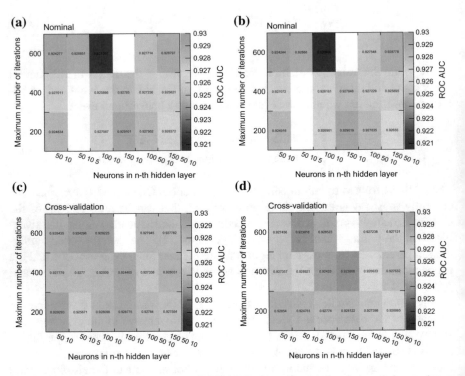

Fig. 14.11 Area under the ROC curve (ROC AUC) depending on the choice of the maximum number of training iterations and the number of neurons per layer for training (left) and testing (right) events. The results of the nominal training and testing phase are shown in the top row, the bottom row shows the result, if the events used for training and testing are swapped. This way, the dependence of the training on the statistics of the input data can be evaluated [23]. Both cases prefer choosing the configuration parameters of 200 iterations and two hidden layers containing 150 and 10 neurons, respectively

Table 14.5 Neural network parameter settings optimised for the search for fully-hadronically decaying top squarks. The optimisation is performed using the four momenta of the four jets leading in p_T including the b-tagging information. Additionally, the missing transverse energy and the azimuthal angle of the missing transverse momentum, $\phi(E_T^{\mathrm{miss}})$ was used in the training. To increase the statistics of signal events in the training, all simulated $\tilde{t}_1 \to t + \tilde{\chi}_1^0$ scenarios are used. The regularisation term is a $L2$ penalty term avoiding large neuron weights to suppress the amount of overtraining [24]

Parameter	Setting
Classifier	MLP
Activation function	ReLU
Number of hidden layers	2
Number of neurons in i-th hidden layer	150, 10
Maximum number of iterations	200
Solver	adam
Learning rate	0.001
β_1	0.95
β_2	0.999
Regularisation term	0.0001

(a)   **(b)**

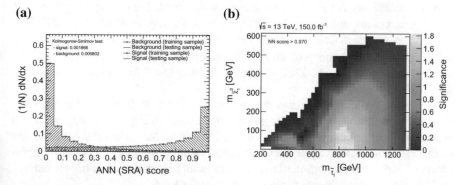

Fig. 14.12 Comparisons of the ANN scores obtained for signal and background in the training and testing phase, respectively, for the ANN targeting SRA (**a**). **b** shows the expected significances as a function of $m_{\tilde{t}_1}$ and $m_{\tilde{\chi}_1^0}$ for a potential SR based on the preselection summarised in Table 14.4 and a requirement on the score of the ANN trained on all signal simulations available for the $\tilde{t}_1 \to t + \tilde{\chi}_1^0$ decay scenario. The results are obtained assuming an integrated luminosity of 150.0 fb^{-1} corresponding to the full LHC Run 2 dataset

the training shows no strong dependency on the settings. Furthermore, it is found that increasing the exponential decay rate for the first moment estimate from $\beta_1 = 0.9$ to $\beta_1 = 0.95$ slightly increases the ROC AUC. The degree of overtraining of the ANN is estimated by the Kolmogorov–Smirnov test for the same two-dimensional scan of the configuration parameters (cf. Fig. E.14 in the Appendix). No sign of overtraining is found for the favoured setting of configuration parameters. The final ANN parameters used are summarised in Table 14.5.

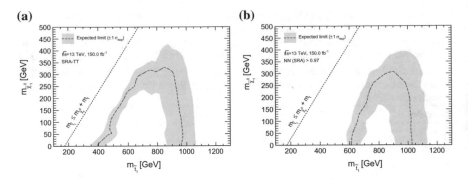

Fig. 14.13 Comparison between the expected exclusion limits at 95% CL for SRA-TT obtained by a simplified maximum likelihood fit assuming an overall systematic uncertainty of 30% and no dedicated CRs (**a**) and a SR based on the preselection summarised in Table 14.4 and a requirement on the score of a NN trained on all signal simulations available for the $\tilde{t}_1 \rightarrow t + \tilde{\chi}_1^0$ decay scenario (**b**). The results are obtained assuming an integrated luminosity of 150.0 fb^{-1} corresponding to the full LHC Run 2 dataset

Figure 14.12a shows the distributions of the ANN score for background and signal events, respectively. No sign for overtraining is observed. The expected significances as a function of $m_{\tilde{t}_1}$ and $m_{\tilde{\chi}_1^0}$ were calculated as previously explained for BDTs. The highest expected significance for high top squark masses and low neutralino masses are achieved for the training on all signal simulations available for the $\tilde{t}_1 \rightarrow t + \tilde{\chi}_1^0$ decay scenario) (cf. Fig. 14.12b). The corresponding cut value on the ANN score is ANN (SRA) > 0.970.

Figure 14.13a shows the expected exclusion limits at 95% CL for SRA-TT obtained with this simplified exclusion fit (also shown in Fig. 14.5a). Figure 14.13b shows the corresponding expected exclusion limits with the same setup but requiring ANN (SRA) > 0.970 on top of the preselection summarised in Table 14.1. The application of the SR exploiting the DNN results in an improvement of approximately 50 GeV (5%) towards higher top squark masses for SRA.

The application of ANNs trained on the jet four momenta and the missing transverse energy yields compatible results as the common SR definitions. This may be seen as an indicator that the usage of ANNs is rather beneficial for applications more close to the reconstruction of actual physics objects than for final discriminating variables.

14.3   Conclusion

Depending on the signal scenario, the application of multivariate techniques used for event classification can result in significant increases of the expected exclusion limits. However, it has to be taken into account that potential problems can arise in the definition of CRs and VRs disjoint to the SRs which are kinematically similar to the SRs using multivariate classifiers. Furthermore, the reduction of several discriminating variables into one single classifier score tremendously limits the understanding of the underlying effect of a single discriminating variable on the event topology. Thus, the application of multivariate techniques has to be carefully considered balancing to its benefits against its drawbacks.

References

1. Friedman JH (2002) Stochastic gradient boosting. Comput Stat Data Anal 38(4):367–378. https://doi.org/10.1016/S0167-9473(01)00065-2
2. Rokach L, Maimon O (2007) Data mining with decision trees: theory and applications. Machine perception and artificial intelligence. World Scientific Pub Co Inc, Singapore. ISBN: 9812771719
3. Schmidhuber J (2014) Deep learning in neural networks: an overview. In: (Apr 2014). arXiv:1404.7828
4. Albertsson K et al (2018) Machine learning in high energy physics community white paper. In: arXiv:1807.02876 [physics.comp-ph]
5. Schapire RE (1990) The strength of weak learnability. Mach Learn 5(2):197–227. https://doi.org/10.1007/BF00116037. ISSN: 1573-0565
6. Freund Y, Schapire RE (1996) Experiments with a new boosting algorithm. In: Proceedings of the 13th international conference on machine learning (ICML'96). Bari, Italy, Morgan Kaufmann Publishers Inc., pp 148–156. ISBN: 1-55860-419-7
7. Freund Y, Schapire RE (1997) A decision-theoretic generalization of on-line learning and an application to boosting. J Comput Syst Sci 55(1):119–139. https://doi.org/10.1006/jcss.1997.1504
8. Kim PJ (1969) On the exact and approximate sampling distribution of the two sample Kolmogorov-Smirnov criterion $D_{mn}, m \leq n$. J Am Stat Assoc 64(328):1625–1637. ISSN: 01621459
9. Eadie WT et al (1971) Statistical methods in experimental physics. American Elsevier Pub Co, Amsterdam
10. Pearson ES, Hartley HO (1976) Biometrika tables for statisticians. Biometrika tables for statisticians Bd. 2. Published for the Biometrika Trustees at the University Press
11. Fawcett T (2006) An introduction to ROC analysis. Pattern Recognit Lett 27(8):861–874. ROC Analysis in Pattern Recognition. https://doi.org/10.1016/j.patrec.2005.10.010. ISSN: 0167-8655
12. Hanley JA, Mcneil B (1982) The meaning and use of the area under a receiver operating characteristic (ROC) Curve. Radiology 143:29–36
13. Graw JM, Köhler NM, Junggeburth JJ (2017) Search for top squark production at the LHC at ps = 13 TeV with the ATLAS detector using multivariate analysis techniques. https://cds.cern.ch/record/2299133
14. Hoecker A et al (2007) TMVA: toolkit for multivariate data analysis. In: PoS ACAT, p 040. arXiv:physics/0703039
15. Pedregosa F et al (2011) Scikit-learn: machine learning in Python. J Mach Learn Res 12:2825–2830
16. Werbos PJ (1975) Beyond regression: new tools for prediction and analysis in the behavioral sciences. Harvard University, Cambridge
17. Rosenblatt F, Cornell Aeronautical Laboratory (1958) The perceptron: a theory of statistical separability in cognitive systems (Project para). Report. Cornell Aeronautical Laboratory, Buffalo

18. Delashmit WH, Manry MT (2005) Recent developments in multilayer perceptron neural networks. In: Proceedings of the 7th annual memphis area engineering and science conference, MAESC
19. Rumelhart DE, McClelland JL, CORPORATE PDP Research Group, eds (1986) Parallel distributed processing: explorations in the microstructure of cognition, vol. 1: Foundations. MIT Press, Cambridge, MA, USA. ISBN: 0-262-68053-X
20. Hastie T, Tibshirani R, Friedman J (2009) The elements of statistical learning: data mining, inference, and prediction, 2nd edn. Springer series in statistics. Springer, Berlin
21. Arora R et al (2016) Understanding deep neural networks with rectified linear units. In: (Nov 2016). arXiv:1611.01491 [cs.LG]
22. Kingma DP, Ba J (2014) Adam: a method for stochastic optimization. In: (Dec 2014). arXiv:1412.6980 [cs.LG]
23. Poole DL, Mackworth AK (2017) Artificial intelligence. Cambridge University Press, Cambridge. ISBN: 9781107195394
24. Goodfellow I, Bengio Y, Courville A (2016) Deep learning. MIT Press, Cambridge. http://www.deeplearningbook.org

Chapter 15
Summary

With the experimental evidence for the existence of Dark Matter and Dark Energy, there is no doubt that the Standard Model (SM) of particle physics is insufficient to fully describe astroparticle physics. Supersymmetry is one of the most favoured theoretical frameworks which gives a particle candidate for Dark Matter and explains the light Higgs boson mass of 125 GeV by the postulation of a light top squark. The latter allows for the direct production of top squarks at the LHC.

A search for the direct production of top squarks decaying into jets and missing transverse energy was performed based on $36.1\,\text{fb}^{-1}$ of proton-proton collision data at a centre-of-mass energy of $\sqrt{s} = 13\,\text{TeV}$ recorded by the ATLAS experiment. Due to the strong selection requirements in terms of large missing transverse energy and high jet multiplicity, the analysis is exploiting phase space regimes where SM predictions based on MC simulations become problematic. In these regions the data-driven estimation of the SM processes is of crucial importance to obtain meaningful SM expectations and control the experimental uncertainties. No significant excess above the Standard Model expectations was observed and exclusion limits at 95% confidence level (CL) were derived. Top squark masses up to 1 TeV are excluded depending on the neutralino mass. The exclusion limits on the neutralino mass can be interpreted as exclusion limits on WIMP production. Sensitivity studies for the search for top squarks using the full LHC Run 2 dataset comprising $150\,\text{fb}^{-1}$ of proton-proton collision data and exploiting multivariate techniques were performed. A 10% improvement in expected sensitivity can be reached employing Boosted Decision Trees trained on the discriminating variables used in the definition of the signal regions of the search. Furthermore, it was found that employing Deep Neural Networks trained only on basic event information, in this case the jet four momenta including the b-tagging information and the missing transverse energy, results in expected sensitivities compatible with the nominal signal region definitions.

© Springer Nature Switzerland AG 2019

N. M. Köhler, *Searches for the Supersymmetric Partner of the Top Quark, Dark Matter and Dark Energy at the ATLAS Experiment*, Springer Theses,
https://doi.org/10.1007/978-3-030-25988-4_15

Based on the same $36.1\,\text{fb}^{-1}$ of proton-proton collision data, a search for WIMP pair production via a spin-0 mediator produced in association with a top quark pair resulting in a similar experimental signature was performed. Mediator masses below $50\,\text{GeV}$ can be excluded at 95% CL assuming a WIMP mass of $1\,\text{GeV}$. The exclusion limits on the WIMP mass are translated into limits on the spin-independent Dark Matter-nucleon scattering cross section which are compared to the results of direct-detection experiments. For WIMP masses below $5\,\text{GeV}$, the results obtained in this search significantly improve the limits on the scattering cross section. However, it has to be taken into account that the results are model-dependent and only valid for the signal scenarios considered in this thesis.

The signal regions sensitive to the search for top squarks were found to be able to constrain a scalar-tensor effective field theory of Dark Energy predicting a light scalar particle coupling to Standard Model particles and possibly being directly produced at the LHC. The first results of a search for Dark Energy performed at a particle accelerator were obtained.

The entirety of physics results published by the ATLAS experiment are heavily relying on a well understood performance of all detector components. The efficiency of the muon reconstruction and identification is measured with a precision at the permille-level and the performance of the Monitored Drift Tube muon chambers under background rates similar to the ones expected for the High Luminosity-LHC is estimated. The measurements based on $4.0\,\text{fb}^{-1}$ of proton-proton collision data recorded in 2016 and 2017 at a centre-of-mass energy of $\sqrt{s} = 13\,\text{TeV}$ are in excellent agreement with previous test beam measurements.

Appendix A
Muon Efficiency Measurements

A.1 Reconstruction Efficiency

Table A.1 lists the selection criteria for the measurement of the muon reconstruction efficiencies. The simulation SM processes used for the evaluation of the MC efficiency as well as for the OC template of the simultaneous fit used for the fake muon background estimation are listed in Table A.2.

A.1.1 Muon Reconstruction Efficiencies and Scale Factors

The reconstruction and identification efficiencies for the *Medium* identification criterion for both data and simulation as well as their ratio, the muon efficiency scale factor, are shown in Fig. A.1. Figures A.2, A.3 and A.4 are showing the same distributions for the *Loose*, *Tight* and *High-p_T* identification criteria.

A.2 Performance of Monitored Drift Tubes

The simulated flux of neutrons in one quadrant of the ATLAS detector at the LHC design luminosity of $1 \cdot 10^{34}\,\mathrm{cm}^{-2}\mathrm{s}^{-1}$ is shown in Fig. A.5. As the simulated flux of photons (cf. 9.1), the innermost chambers of the inner end-cap layer (EI) are permeated by the highest neutron flux.

Table A.3 summarises the dataset used for studying the MDT background hit rates, spatial resolution and chamber efficiencies presented in Chap. 9. The dataset covers a range of instantaneous luminosities from $\mathcal{L} = 0.15 \cdot 10^{33}\,\mathrm{cm}^{-2}\mathrm{s}^{-1}$ up to $\mathcal{L} = 20.61 \cdot 10^{33}\,\mathrm{cm}^{-2}\mathrm{s}^{-1}$ while containing $3.95\,\mathrm{fb}^{-1}$ of pp collision data.

Table A.4 summarises the selection criteria applied for the measurement of the MDT chamber reconstruction efficiency as well as for the estimation of the spatial

© Springer Nature Switzerland AG 2019
N. M. Köhler, *Searches for the Supersymmetric Partner of the Top Quark, Dark Matter and Dark Energy at the ATLAS Experiment*, Springer Theses, https://doi.org/10.1007/978-3-030-25988-4

Table A.1 Selection criteria applied for the measurement of the muon reconstruction efficiencies. $\mathscr{L}$ (10^{34} cm^{-2}s^{-1}) is denoting the instantaneous luminosity

Tag selection		
Trigger	Unprescaled single muon trigger HLT threshold: 21 GeV, L1 threshold: 15 GeV (2015) HLT threshold: 25.2 GeV, L1 threshold: 20 GeV (2016, $\mathscr{L} \leq 1.02$) HLT threshold: 27.3 GeV, L1 threshold: 20 GeV (2016, $\mathscr{L} > 1.02$)	
Kinematics isolation	$p_T > 28$ GeV, $|\eta| < 2.5$ requirements on $p_T^{\text{varcone30}}/p_T$ and E_T^{cone20}/E_T keeping isolation efficiency constant at 99% in η and p_T	
Probe selection		
Kinematics isolation	$p_T > 10$ GeV, $|\eta| < 2.5$ requirements on $p_T^{\text{varcone30}}/p_T$ and E_T^{cone20}/E_T keeping isolation efficiency constant at 99% in η and p_T	
	Calorimeter-tagged muons	ME tracks
Track requirements	ID track quality cuts mentioned in Sect. 7.2	–
Invariant mass	$|m_{\text{tag-probe}} - m_Z| < 10$ GeV	
Electric charge	$q_{\text{tag}} \cdot q_{\text{probe}} < 0$	

Table A.2 Overview of the simulations used for the SM processes in the $Z \rightarrow \mu^+\mu^-$ tag-and-probe efficiency measurements. More details of the generator configurations can be found in [1–4]. For the simulation of diboson production, $WW \rightarrow \ell\nu\ell\nu$, $WZ \rightarrow \ell\nu\ell\ell$, $ZZ \rightarrow \ell\ell\ell\ell$, $ZZ \rightarrow \nu\nu\ell\ell$, $WZ \rightarrow qq\ell\ell$, $ZZ \rightarrow qq\ell\ell$ processes are considered

Process	Generator	Showering	PDF set	UE tune	Order
$Z \rightarrow \mu^+\mu^-$	Powheg- Box 2	Pythia 8.186	CTEQ6L1	AZNLO	NLO
Diboson	Powheg- Box 2	Pythia 8.186	CTEQ6L1	AZNLO	NLO
$Z \rightarrow \tau^+\tau^-$	Powheg- Box 2	Pythia 8.186	CTEQ6L1	AZNLO	NLO
$t\bar{t}$	Powheg- Box 2	Pythia 8.186	NNPDF2.3LO	A14	LO
$W(\rightarrow \mu\nu)$+jets	Powheg- Box 2	Pythia 8.186	CTEQ6L1	AZNLO	NLO
$b\bar{b}$	PYTHIA8B	PYTHIA8B	NNPDF2.3LO	A14	LO
$c\bar{c}$	PYTHIA8B	PYTHIA8B	NNPDF2.3LO	A14	LO

resolution. Figure A.6a, b show the uncertainty on the spatial resolution depending on the instantaneous luminosity for the MDT chambers located in the innermost end-cap (EI) and the middle enc-cap (EM) layers, respectively. Figure A.7 shows the uncertainty on the spatial resolution depending on the instantaneous luminosity for the MDT chambers located in the innermost barrel (BI) layer for $|\eta| > 0$.

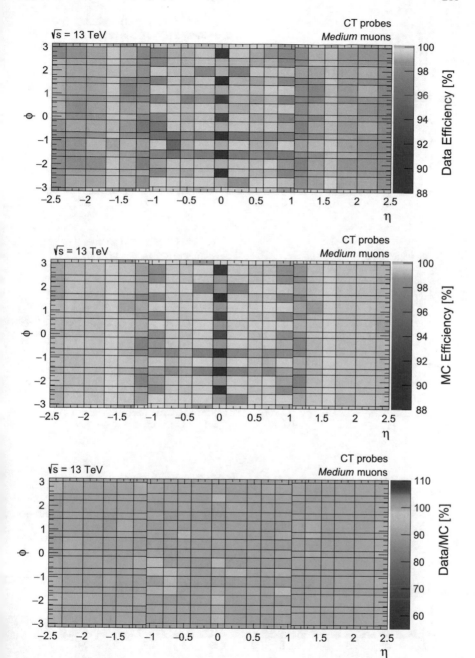

Fig. A.1 Muon reconstruction efficiency for the *Medium* identification algorithm measured in 2016 data, MC simulation as well as the efficiency scale factor as a function of the muon η and ϕ for muons with $p_T > 10$ GeV. The binning reflects the geometry of the ATLAS muon spectrometer

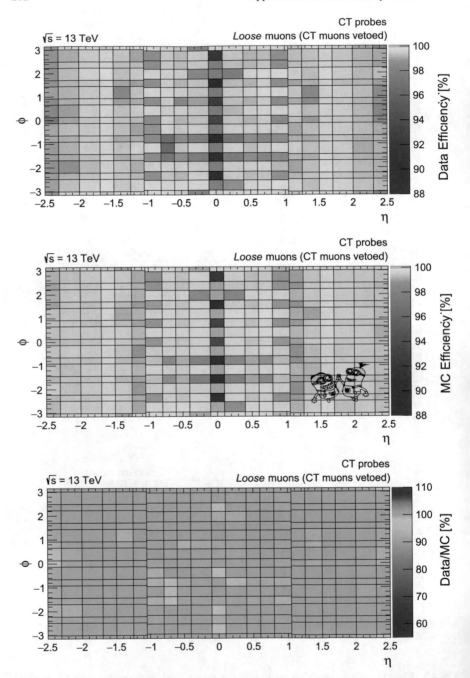

Fig. A.2 Muon reconstruction efficiency for the *Loose* identification algorithm measured in 2016 data, MC simulation as well as the efficiency scale factor as a function of the muon η and ϕ for muons with $p_T > 10$ GeV. The efficiency only reflects muons satisfying the *Loose* identification algorithm but not being calorimeter-tagged muons. The binning reflects the geometry of the ATLAS muon spectrometer

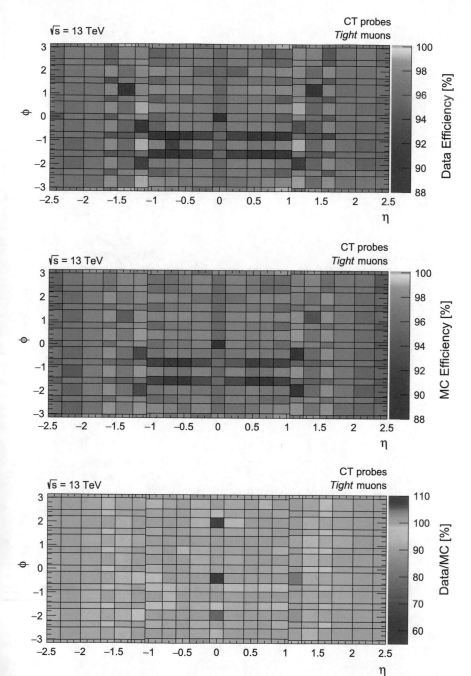

Fig. A.3 Muon reconstruction efficiency for the *Tight* identification algorithm measured in 2016 data, MC simulation as well as the efficiency scale factor as a function of the muon η and ϕ for muons with $p_T > 10$ GeV. The binning reflects the geometry of the ATLAS muon spectrometer

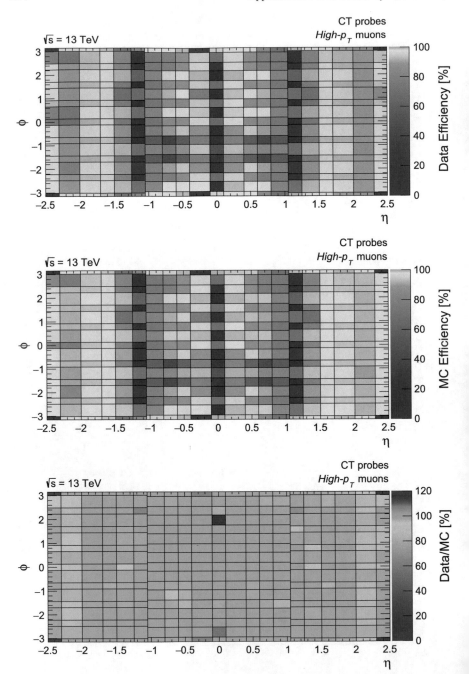

Fig. A.4 Muon reconstruction efficiency for the *High-p$_T$* identification algorithm measured in 2016 data, MC simulation as well as the efficiency scale factor as a function of the muon η and ϕ for muons with $p_T > 30$ GeV. The binning reflects the geometry of the ATLAS muon spectrometer

Fig. A.5 Simulated flux of neutrons in one quadrant of the ATLAS detector at the LHC design luminosity of $1 \cdot 10^{34}$ cm^{-2}s^{-1}. The inner (I), middle (M) and outer (O) layers of muon chambers in barrel (B) and end-caps (E) are indicated [5]

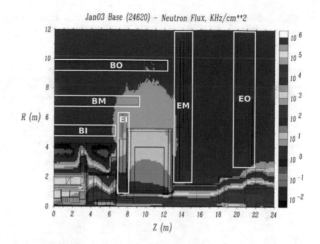

Table A.3 Dataset used for studying the behaviour of ATLAS MDT chambers depending on the instantaneous luminosity

Recording date	Instantaneous luminosity $\mathscr{L}$ (10^{33} cm^{-2}s^{-1})	Integrated luminosity L (fb^{-1})
2016, April 28th	0.150–0.198	0.003
2016, May 12th	1.530–1.969	0.0445
2016, May 28th	3.343–4.562	0.0944
2016, June 2nd	4.374–7.476	0.334
2016, October 2nd	5.001–12.658	0.348
2016, October 9th	11.064–13.171	0.030
2017, June 4th	1.757–3.216	0.076
2017, June 5th	2.191–2.922	0.040
2017, July 20th	5.705–14.860	0.460
2017, August 9th	13.974–17.466	0.156
2017, August 18th	9.176–10.534	0.070
2017, September 8th	5.647–10.872	0.325
2017, September 26th	7.280–14.019	0.399
2017, October 4th	6.710–17.078	0.409
2017, October 16th	6.917–15.696	0.462
2017, October 17th	6.795–20.486	0.243
2017, November 2nd	6.151–20.614	0.456
Total	0.150–20.614	3.950

Table A.4 Selection criteria applied for the measurement of the MDT chamber reconstruction efficiency. The same muon selection is also applied for the estimation of the spatial resolution depending on the background hit rates

	Requirement
Trigger	No trigger requirement
Muon selection	At least one muon
	Medium identification criterion
	$p_T > 20\,\text{GeV}$, $|\eta| < 2.5$
Veto cosmic muons	$|z_0| < 1\,\text{mm}$, $|d_0| < 0.2\,\text{mm}$
Track-to-vertex association	$|z_0| \cdot \sin(\theta) < 0.5\,\text{mm}$, $|d_0|/\sigma_{d_0} < 3$
Efficiency estimation	CT muon extrapolated to chamber
	Medium muon with ≥ 1 hit in chamber associated to the muon track
	Event not recorded because of *Medium* muon
	Matching: $\Delta R(\text{CT}, \mu) < 0.05$

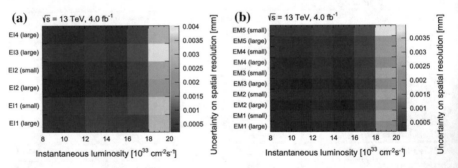

Fig. A.6 Uncertainty on the spatial resolution depending on the instantaneous luminosity (cf. Fig. 9.5) for the different layers of EI (**a**) and EM (**b**) MDT chambers. The measurement combines the information of all MDT chambers of the same type for a given layer. EI1–EI4 (EM1–EM5) denotes the distance of the chamber layer with respect to the beam pipe. Small and large chambers are shown separately

A.2.1 Estimation of MDT Chamber Efficiency

As described in Sect. 9.3, the chamber efficiency is estimated by extrapolating calorimeter-tagged muons to the chamber of interest and searching reconstructed muons passing the *Medium* identification criterion and having at least one hit associated to the muon track which has to be located within $\Delta R < 0.05$ from the calorimeter-tagged muon. Applying the selection requirements summarised in Table A.4, the chamber efficiency shown in Fig. A.8a is obtained. Compared to the muon reconstruction efficiency of the *Medium* identification criterion measured in data (cf. Fig. A.1), an efficiency loss of approximately 10% is observed (cf. Fig. A.8a). The efficiency drop arises from two independent effects:

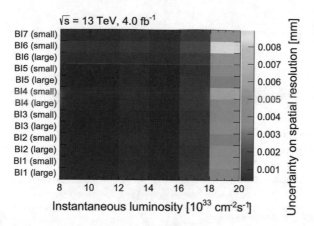

Fig. A.7 Uncertainty on the spatial resolution depending on the instantaneous luminosity for the different layers of the innermost layer of barrel chambers (BI). The measurement combines the information of all MDT chambers of the same type for a given layer in $\eta > 0$. BI1–BI7 denotes the distance of the barrel chamber layer with respect to the transverse plane. Small and large chambers are shown separately

- The *Medium* identification criterion for muons within $0.1 < |\eta| < 2.5$ requires the presence of at least three MDT hits in at least two chamber layers (cf. Chap. 7) in order to obtain a meaningful result in the track fit and suppress the contribution of fake muons. The measurement of the chamber efficiency does not apply this requirement which makes it possible that fake calorimeter-tagged muons are extrapolated to the chamber of interest, but no muon track can be associated to them. Requiring that at least one MDT in the middle end-cap chamber traversed by a muon satisfying the *Medium* identification criterion is associated to the muon track increases the chamber efficiency by approximately 3% (cf. Fig. A.8b).
- The extrapolation of the calorimeter-tagged muons to the chamber of interest exploits the total surface of the MDT chamber. In case the track of the calorimeter-tagged muon is only traversing the edges of the chamber, it still contributes to the efficiency measurement resulting in an efficiency loss if no hit is recorded. Tightening the angular requirements in order to mask the chamber boundaries (cf. Fig. A.8c), increases the chamber efficiency to values comparable with the reconstruction efficiency measurement.

Figure A.8d shows the comparison between the relative change of the chamber efficiency depending on the instantaneous luminosity for the three definitions of the chamber efficiencies discussed above. The efficiency is normalised to the lowest bin of the instantaneous luminosity. Within statistical uncertainties, the relative change of the chamber efficiency agree within the three chamber efficiency definitions within two standard deviations. Thus, the relative change of the chamber efficiency measured with the selection described in Sect. 9.3 normalised to the lowest bin in instantaneous luminosity is used as a measure for the dependence of the chamber efficiency on the background hit rates.

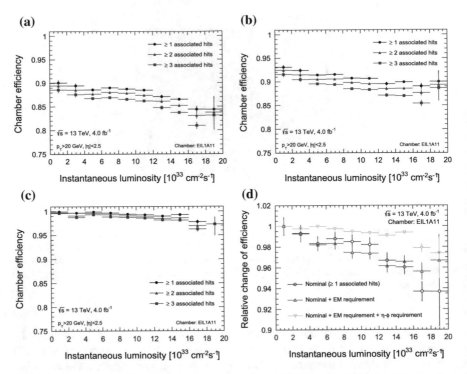

Fig. A.8 Measurement of the chamber efficiency of one of the innermost EI chambers based on the selection criteria summarised in Table A.4 (**a**), with an additional requirement on the presence of an MDT hit in the middle end-cap layer (**b**) and a tighter angular selection masking the chamber boundaries (**c**). **d** Shows the relative change of the chamber efficiency normalised to the lowest bin in instantaneous luminosity for scenarios (**a**)–(**c**)

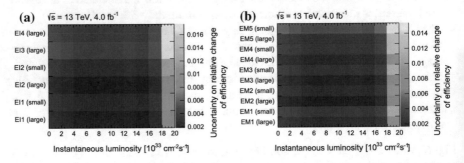

Fig. A.9 Uncertainties on the relative change of chamber efficiencies (cf. Fig. 9.11) depending on instantaneous luminosity for the different layers of EI (**a**) and EM (**b**) MDT chambers. The measurement combines the information of all MDT chambers of the same type for a given layer. EI1–EI4 (EM1–EM5) denotes the distance of the chamber layer with respect to the beam pipe. Small and large chambers are shown separately

Figure A.9 shows the uncertainties on the relative change of the chamber efficiencies depicted in Fig. 9.11. As expected, the smaller chambers show slightly larger statistical uncertainties.

References

1. ATLAS Collaboration (2016) Simulation of top-quark production for the ATLAS experiment at $\sqrt{s} = 13 TeV$. ATL-PHYS-PUB-2016-004. https://cds.cern.ch/record/2120417
2. ATLAS Collaboration (2016) Monte Carlo generators for the production of a W or $Z/\gamma*$ Boson in association with Jets at ATLAS in run 2. ATL-PHYS-PUB-2016-003. https://cds.cern.ch/record/2120133
3. ATLAS Collaboration (2016) Multi-boson simulation for 13 TeV ATLAS analyses. ATL-PHYS-PUB-2016-002. https://cds.cern.ch/record/2119986
4. ATLAS Collaboration (2016) Modelling of the $t\bar{t}H$ and $t\bar{t}V(V = W, Z)$ processes for $\sqrt{s} = 13 TeV$ ATLAS analyses. ATL-PHYS-PUB-2016-005. https://cds.cern.ch/record/2120826
5. Schwegler P, Kroha H (2014) High-rate performance of muon drift tube detectors. Presented 14 July 2014. https://cds.cern.ch/record/1746370

Appendix B
Search for Direct Top Squark Pair Production

B.1 Estimation of Missing Transverse Energy Trigger Efficiencies

Since the E_T^{miss} is an observable calculated from several objects in the event, the efficiency of the E_T^{miss} trigger is strongly dependant on the event selection for a given analysis. A selection similar to the one of the W boson cross section measurement [1] is applied, namely requiring exactly one muon with $p_T > 20\,\text{GeV}$, after a single muon trigger has triggered the TDAQ, with a transverse mass between the muon and the E_T^{miss} of $m_T\left(\mu, E_T^{miss}\right) > 50\,\text{GeV}$. Analysis specific requirements such as the presence of at least four jets and two b-jets are additionally added. The full list of requirements is summarised in Table B.1. The efficiency of a E_T^{miss} trigger is calculated by the number of all events passing the selection described in Table B.1 and passing the E_T^{miss} trigger, divided by all events passing the selection. The efficiencies for the lowest unprescaled E_T^{miss} triggers used in 2015 and 2016 are shown in Fig. B.1 as a function of the E_T^{miss} calculated with no contributions in order to emulate the calorimeter-based E_T^{miss} calculation only. The E_T^{miss} trigger with an HLT threshold of $110\,\text{GeV}$ becomes more than 99% efficient for a $E_T^{miss} > 250\,\text{GeV}$ requirement.

B.2 Validation of Simulated Signal Processes Using the Fast Simulation Framework

B.2.1 Top Squark Decays into Top Quark and Neutralino

In order to ensure that the usage of simulations where the showers in the electromagnetic and hadronic calorimeters are simulated with a parametrised description instead of using GEANT 4 does not affect the distributions of discriminating variables, for some signal scenarios the simulated events are also reconstructed using

© Springer Nature Switzerland AG 2019
N. M. Köhler, *Searches for the Supersymmetric Partner of the Top Quark, Dark Matter and Dark Energy at the ATLAS Experiment*, Springer Theses, https://doi.org/10.1007/978-3-030-25988-4

Table B.1 Selection requirements for the estimation of the efficiency of the E_T^{miss} triggers

	Requirement
Trigger	Single muon triggers as described in Table 11.9
Number of jets	$N_{jets} \geq 4$
Number of b-jets	$N_{b-jets} \geq 2$
Number of muons	$N_\mu = 1$
Muon p_T	$p_{T,\mu}^0 > 20\,\text{GeV}$
Transverse mass between muon and E_T^{miss}	$m_T\left(\mu, E_T^{miss}\right) > 50\,\text{GeV}$

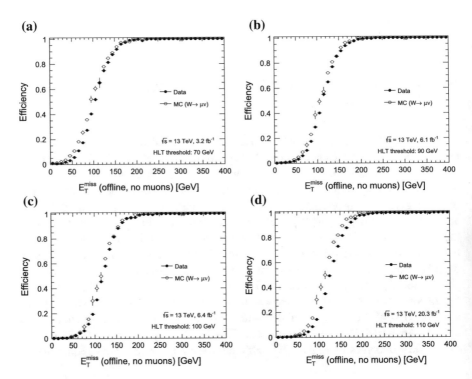

Fig. B.1 E_T^{miss} trigger efficiency curves with respect to the E_T^{miss} reconstructed offline without muon corrections for all events passing a $W \rightarrow \mu\nu$ selection for different trigger objects and data-taking periods

GEANT 4 for the calorimeter showers. Figures B.2, B.3 and B.4 show a selection of simulated distributions of discriminating variables used in the search for top squarks normalised to an integrated luminosity of $1\,\text{fb}^{-1}$. The Figures compare the simulation reconstructed with the parametrised description of the calorimeter showers (fast simulation) with the full GEANT 4 setup. All distributions are in good agreement within the statistical and experimental systematic uncertainties.

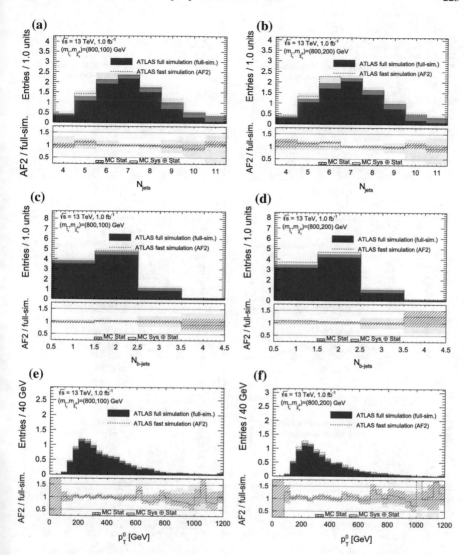

Fig. B.2 Comparison between the full ATLAS detector simulation based on GEANT 4 (red) and the fast simulation framework (AF2) for top squark pair production with $m_{\tilde{t}} = 800$ GeV and $m_{\tilde{\chi}_1^0} = 100$ GeV (left) or $m_{\tilde{\chi}_1^0} = 200$ GeV (right), respectively. The hashed (yellow) bands are indicating the statistical (experimental systematic) uncertainties

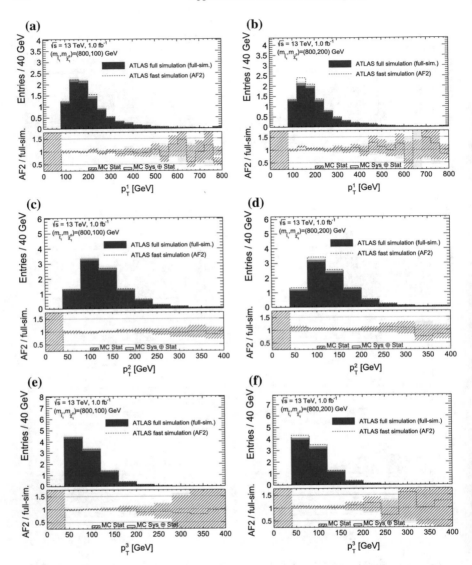

Fig. B.3 Comparison between the full ATLAS detector simulation based on GEANT 4 (red) and the fast simulation framework (AF2) for top squark pair production with $m_{\tilde{t}} = 800$ GeV and $m_{\tilde{\chi}_1^0} = 100$ GeV (left) or $m_{\tilde{\chi}_1^0} = 200$ GeV (right), respectively. The hashed (yellow) bands are indicating the statistical (experimental systematic) uncertainties

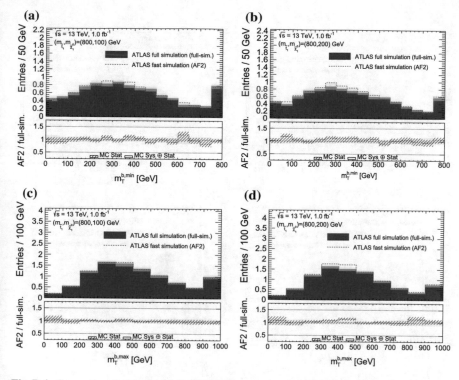

Fig. B.4 Comparison between the full ATLAS detector simulation based on GEANT 4 (red) and the fast simulation framework (AF2) for top squark pair production with $m_{\tilde{t}} = 800$ GeV and $m_{\tilde{\chi}_1^0} = 100$ GeV (left) or $m_{\tilde{\chi}_1^0} = 200$ GeV (right), respectively. The hashed (yellow) bands are indicating the statistical (experimental systematic) uncertainties

B.2.2 Top Squark Decays Involving Charginos

Figures B.5 and B.6 show a selection of simulated distributions of discriminating variables used in the search for top squarks normalised to an integrated luminosity of $1\,\text{fb}^{-1}$ for scenarios involving charginos in the decay chain. As for the signal scenario with $\mathcal{BR}(\tilde{t}_1 \rightarrow t + \tilde{\chi}_1^0) = 100\%$, all distributions are in good agreement within the statistical and experimental systematic uncertainties.

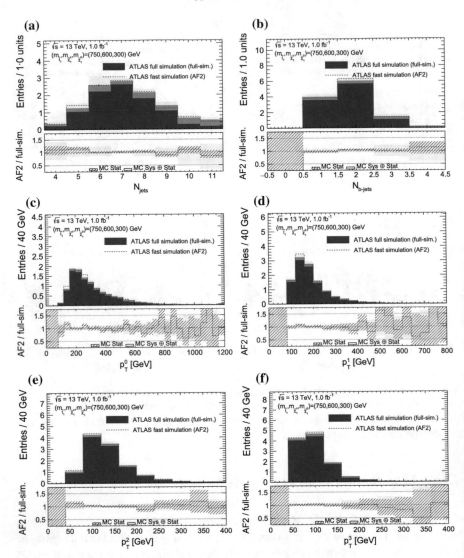

Fig. B.5 Comparison between the full ATLAS detector simulation based on GEANT 4 (red) and the fast simulation framework (AF2) for top squark pair production with subsequent decays involving charginos. The hashed (yellow) bands are indicating the statistical (experimental systematic) uncertainties

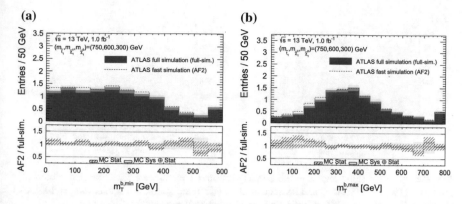

Fig. B.6 Comparison between the full ATLAS detector simulation based on GEANT 4 (red) and the fast simulation framework (AF2) for top squark pair production with subsequent decays involving charginos. The hashed (yellow) bands are indicating the statistical (experimental systematic) uncertainties

B.3 Optimisation of Signal Regions Targeting Large Top Squark Masses

In the optimisation for SRA-TT, requirements on $E_T^{miss} > 400$ GeV and $m_{T2} > 400$ GeV are imposed in addition to the basic requirements summarised in Table 11.1 and the TT category requirements on the $R = 1.2$ reclustered jet masses (cf. Fig. B.8a). SRA-TW and SRA-T0 both impose a requirement on $m_{jet, R=0.8}^0 > 60$ GeV (cf. Fig. B.8b). The simulated distribution of $m_{jet, R=0.8}^0$ is shown in Fig. B.7. Requiring $m_{jet, R=0.8}^0 > 60$ GeV in SRA-TT as well does not result in any loss of expected significance.

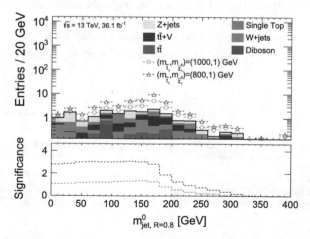

Fig. B.7 Simulated distribution of $m_{jet, R=0.8}^0$ after requiring $E_T^{miss} > 400$ GeV and $m_{T2} > 400$ GeV on top of the selection mentioned in Fig. 11.13 for the TT category of SRA. The bottom panel shows the expected significance for the signal scenarios with $(m_{\tilde{t}_1}, m_{\tilde{\chi}_1^0}) = (1000, 1)$ GeV and $(m_{\tilde{t}_1}, m_{\tilde{\chi}_1^0}) = (800, 1)$ GeV depending on a potential cut value on the corresponding variable

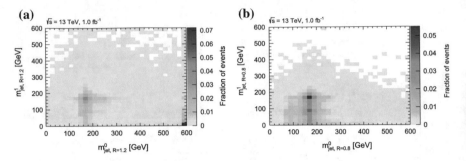

Fig. B.8 $R = 1.2$ (left) and $R = 0.8$ (right) reclustered jet masses for simulated gluino pair production ($m_{\tilde{g}} = 1700$ GeV) and subsequent decays into top quark, top squark ($m_{\tilde{t}_1} = 400$ GeV) and neutralino ($m_{\tilde{\chi}_1^0} = 395$ GeV) after the preselection requirements (cf. Table 11.1)

B.4 Pair Production of Gluinos with Subsequent Top Squark Decays

In order to suppress SM $t\bar{t}$ background, the $m_T^{b,\min} > 200$ GeV requirement is applied for the optimisation of the selection targeting the pair production of gluinos with subsequent top squark decays. Figure B.9a shows the simulated distribution of E_T^{miss} after applying the basic requirements listed in Table 11.1 as well as $m_T^{b,\min} > 200$ GeV. Although the expected significance for the signal scenario with $(m_{\tilde{g}}, m_{\tilde{t}_1}, m_{\tilde{\chi}_1^0}) = (1700, 400, 395)$ GeV is increasing with a potential cut value on E_T^{miss}, in order to preserver enough statistics in the SR, $E_T^{\text{miss}} > 550$ GeV is required since it does not decrease any expected significance. The same arguments holds for requiring $E_T^{\text{miss}}/\sqrt{H_T} > 18$ (cf. Fig. B.9b). Analogously to the categories in SRA exploiting the $R = 1.2$ reclustered jets, for the gluino pair production, the decay objects are expected to be even more boosted which allows to require $m_{\text{jet}, R=0.8}^0 > 120$ GeV and $m_{\text{jet}, R=0.8}^1 > 80$ GeV. The simulated distribution of H_T is shown in Fig. B.10a

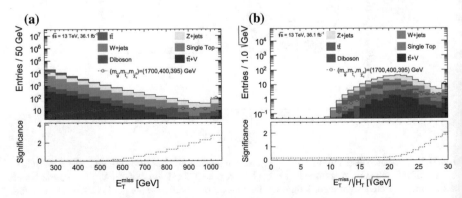

Fig. B.9 With $m_T^{b,\min} > 200$ GeV (**a**) and $E_T^{\text{miss}} > 550$ GeV (**b**)

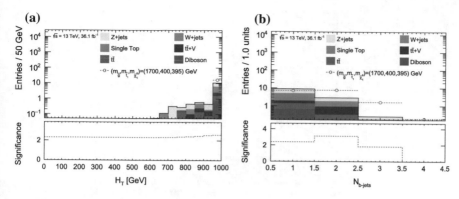

Fig. B.10 With all requirements from Table B.2 except for H_T and $N_{b-\mathrm{jets}}$

Table B.2 Signal region selection targeting the pair production of gluinos with subsequent top squark decays in addition to the requirements presented in Table 11.1

Variable	SRE
$\left\|\Delta\phi\left(\mathrm{jet}^{0,1,2}, E_T^{\mathrm{miss}}\right)\right\|$	>0.4
b-tagged jets	≥ 2
$m_T^{b,\mathrm{min}}$	$>200\,\mathrm{GeV}$
E_T^{miss}	$>550\,\mathrm{GeV}$
$m_{\mathrm{jet},R=0.8}^{0}$	$>120\,\mathrm{GeV}$
$m_{\mathrm{jet},R=0.8}^{1}$	$>80\,\mathrm{GeV}$
H_T	$>800\,\mathrm{GeV}$
$E_T^{\mathrm{miss}}/\sqrt{H_T}$	$>18\,\sqrt{\mathrm{GeV}}$

after applying all requirements discussed. The expected significance in the lower panel does not change up to $H_T > 800\,\mathrm{GeV}$ which allows to add this requirement to further suppress SM backgrounds. Subsequently, the simulated number of b-jets is shown in Fig. B.10b with applying the requirement on H_T. Requiring the presence of at least two b-jets further increases the expected sensitivity. All requirements for the SR targeting the pair production of gluinos with subsequent decays into top squarks can be found in Table B.2.

Figure B.11 shows the ratio between E_T^{miss} and the square root of H_T after applying all cuts of SRE except the cut on $E_T^{\mathrm{miss}}/\sqrt{H_T}$ itself. The most dominant SM contributions are arising from Z+jets and $t\bar{t}$ production. The expected yields in SRE (cf. Table B.2) for an integrated luminosity of $36.1\,\mathrm{fb}^{-1}$ are shown in Table B.3.

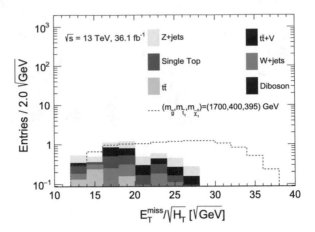

Fig. B.11 Distribution of $E_T^{miss}/\sqrt{H_T}$ after applying all requirements of SRE (cf. Table B.2) but the one on $E_T^{miss}/\sqrt{H_T}$ scaled to an integrated luminosity of 36.1 fb^{-1} without drawing statistical or systematic uncertainty bands. In addition to the SM processes, also simulated gluino pair production with $(m_{\tilde{g}}, m_{\tilde{t}_1}, m_{\tilde{\chi}_1^0}) = (1700, 400, 395)$ GeV is shown. The rightmost bin includes overflow events

B.5 Definition of Control Regions for Search for Top Squark Pair Production

Top Quark Pair Production in Association with a Vector Boson

Figure B.12 shows the relative SR contributions of $t\bar{t} + Z$ and $t\bar{t} + W$ production with respect to $t\bar{t} + V$ shown in Table B.3. For SRE, more than 86% of the $t\bar{t} + V$ contribution is arising from $t\bar{t} + Z$ production.

Table B.3 Expected number of simulated events in the signal region targeting gluino pair production with subsequent decays involving top squarks scaled to an integrated luminosity of 36.1 fb^{-1}. For the signal scenario used for the optimisation of the SR, the expected significance is given. For the SM processes, the statistical and the experimental systematic uncertainties are shown

	SRE
$(m_{\tilde{g}}, m_{\tilde{t}_1}, m_{\tilde{\chi}_1^0}) = (1700, 400, 395)$ GeV	$9.49 \pm 0.15 \ (3.1\sigma)$
$t\bar{t}$	$0.21 \pm 0.06^{+0.075}_{-0.069}$
Z+jets	$1.15 \pm 0.18^{+0.227}_{-0.210}$
W+jets	$0.42 \pm 0.13^{+0.159}_{-0.270}$
Single top	$0.56 \pm 0.14^{+0.072}_{-0.066}$
$t\bar{t} + V$	$0.69 \pm 0.13^{+0.137}_{-0.145}$
Diboson	$0.14 \pm 0.08 \pm 0.041$
Total SM	$3.17 \pm 0.31^{+0.411}_{-0.531}$

Fig. B.12 Relative contributions of $t\bar{t} + Z$ and $t\bar{t} + W$ production in all SRs exploited in the search for top squarks with fully-hadronic final states

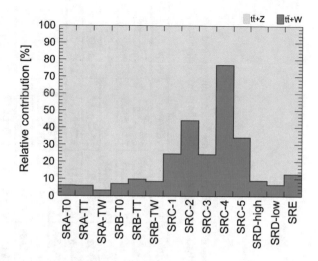

Table B.4 Selection requirements for the comparison of the transverse momenta of the Z boson (photon) arising from $t\bar{t} + Z$ ($t\bar{t} + \gamma$) production at generator-level

Variable	$t\bar{t} + Z$	$t\bar{t} + \gamma$
Number of jets	$N_{\text{jets}} \geq 4$	
Number of b-jets	$N_{b-\text{jets}} \geq 2$	
Number of leptons	$N_\ell = 0$	$N_\ell = 1$
Number of photons	$N_\gamma = 0$	$N_\gamma = 1$
Lepton p_T	–	$p_{T,\ell}^0 > 28\,\text{GeV}$
Photon p_T	–	$p_{T,\gamma}^0 > 150\,\text{GeV}$
Z boson p_T	$p_{T,Z}^0 > 150\,\text{GeV}$	–

In order to estimate the contribution of $t\bar{t} + Z(\to \nu\nu)$ in the SRs targeting top squark decays, a $t\bar{t} + \gamma$ CR is defined. For validating that $t\bar{t} + \gamma$ production is describing the same event topology as $t\bar{t} + Z(\to \nu\nu)$ for large transverse momenta of the Z/γ^* boson, a comparison at generator-level is performed applying the requirements listed in Table B.4.

B.6 Definition of Validation Regions for Search for Top Squark Pair Production

This section lists the detailed selection criteria for all VRs exploited in the search for top squarks.

The composition of the different SM processes as well as the observed events in pp collision data in all VRs validating the transfer factors for SM $t\bar{t}$ production is given in Table B.5.

Table B.5 Composition of SM processes in $t\bar{t}$ validation regions for 36.1 fb^{-1} of pp collision data. None of the simulated SM predictions is normalised using normalisation factors derived from a simultaneous fit (detailed in Sect. 11.6). The purities of the simulated $t\bar{t}$ contributions as well as the normalisation factors computed with Eq. (10.1) are also shown for all regions

	VRTA-TT	VRTA-TW	VRTA-T0	VRTB-TT	VRTB-TW	VRTB-T0	VRTC	VRTD	VRTE
$t\bar{t}$	30.92 ± 1.11	21.28 ± 0.73	39.24 ± 0.95	51.08 ± 1.71	61.14 ± 1.97	112.81 ± 3.12	215.6 ± 3.44	147.36 ± 2.89	65.57 ± 2.02
Z+jets	2.67 ± 0.3	2.63 ± 0.32	5.16 ± 0.63	4.07 ± 0.39	5.7 ± 0.62	17.98 ± 1.57	30.03 ± 1.15	7.56 ± 0.62	3.58 ± 0.38
W+jets	0.9 ± 0.22	0.78 ± 0.21	1.8 ± 0.45	1.43 ± 0.26	2.87 ± 0.95	3.55 ± 0.77	36.11 ± 2.18	3.19 ± 0.66	1.94 ± 0.4
Single top	1.84 ± 0.38	0.62 ± 0.13	1.65 ± 0.37	3.14 ± 0.67	3.63 ± 0.76	8.08 ± 1.11	24.29 ± 3.5	10.43 ± 1.08	3.8 ± 0.67
$t\bar{t}+V$	1.79 ± 0.26	1.51 ± 0.2	1.73 ± 0.22	3.48 ± 0.36	3.23 ± 0.32	5.26 ± 0.36	6.82 ± 0.57	3.77 ± 0.34	2.28 ± 0.33
Diboson	0.16 ± 0.1	0.69 ± 0.42	0.24 ± 0.11	0.15 ± 0.1	0.6 ± 0.23	1.82 ± 0.67	4.65 ± 0.99	0.56 ± 0.17	0.48 ± 0.31
Total MC	38.28 ± 1.26	27.5 ± 0.95	49.81 ± 1.31	63.36 ± 1.93	77.17 ± 2.43	149.5 ± 3.82	317.49 ± 5.61	172.86 ± 3.24	77.65 ± 2.24
Data	46.0 ± 6.78	35.0 ± 5.92	56.0 ± 7.48	83.0 ± 9.11	106.0 ± 10.3	164.0 ± 12.81	277.0 ± 16.64	215.0 ± 14.66	105.0 ± 10.25
Purity of $t\bar{t}$	0.81 ± 0.04	0.77 ± 0.04	0.79 ± 0.03	0.81 ± 0.04	0.79 ± 0.04	0.75 ± 0.03	0.68 ± 0.02	0.85 ± 0.02	0.84 ± 0.04
Normalisation of $t\bar{t}$	1.25 ± 0.23	1.35 ± 0.29	1.16 ± 0.2	1.38 ± 0.19	1.47 ± 0.18	1.13 ± 0.13	0.81 ± 0.09	1.29 ± 0.11	1.42 ± 0.17

The composition of the different SM processes as well as the observed events in pp collision data in all VRs validating the transfer factors for SM Z+jets production is given in Table B.6.

The composition of the different SM processes as well as the observed events in pp collision data in the VR validating the transfer factor for SM W+jets production is given in Table B.7.

The number of both observed and predicted event yields after the likelihood fit for all VRs are shown in Fig. B.13.

Table B.6 Composition of SM processes in Z+jets validation regions for 36.1 fb^{-1} of pp collision data. None of the simulated SM predictions is normalised using normalisation factors derived from a simultaneous fit (detailed in Sect. 11.6). The purities of the simulated Z+jets contributions as well as the normalisation factors computed with Eq. (10.1) are also shown for all regions

	VRZAB	VRZD	VRZE
$t\bar{t}$	17.2 ± 1.44	48.19 ± 2.04	13.36 ± 0.73
Z+jets	56.97 ± 2.15	93.77 ± 2.86	34.42 ± 1.71
W+jets	31.31 ± 2.44	39.63 ± 2.52	19.45 ± 1.7
Single top	4.46 ± 0.59	7.92 ± 0.81	8.34 ± 0.72
$t\bar{t} + V$	2.14 ± 0.22	5.97 ± 0.41	2.81 ± 0.28
Diboson	8.59 ± 1.64	12.66 ± 1.78	5.69 ± 1.25
Total MC	120.67 ± 3.97	208.14 ± 4.76	84.08 ± 2.92
Data	142.0 ± 11.92	254.0 ± 15.94	112.0 ± 10.58
Purity of Z+jets	0.47 ± 0.02	0.45 ± 0.02	0.41 ± 0.02
Normalisation of Z+jets	1.37 ± 0.24	1.49 ± 0.19	1.81 ± 0.34

Table B.7 Composition of SM processes in the W+jets validation region for 36.1 fb^{-1} of pp collision data. None of the simulated SM predictions is normalised using normalisation factors derived from a simultaneous fit (detailed in Sect. 11.6). The purity of the simulated W+jets contribution as well as the normalisation factor computed with Eq. (10.1) are also shown

	VRW
$t\bar{t}$	267.68 ± 4.24
Z+jets	3.47 ± 0.47
W+jets	475.77 ± 10.79
Single Top	123.3 ± 3.98
$t\bar{t} + V$	2.04 ± 0.21
Diboson	19.44 ± 2.54
Total MC	891.7 ± 12.53
Data	985.0 ± 31.38
Purity of W+jets	0.53 ± 0.01
Normalisation of W+jets	1.2 ± 0.08

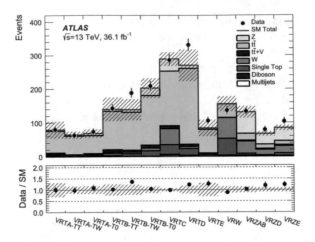

Fig. B.13 Observed and predicted event yields for all validation regions after the likelihood fit. The stacked histograms show the SM prediction and the hatched uncertainty band around the SM prediction shows the total uncertainty, which consists of the MC statistical uncertainties, detector-related systematic uncertainties, and theoretical uncertainties in the extrapolation from CR to VR [2]

B.7   Estimation of Multi-jet Production

This sections lists the detailed selections applied to define CRs and VRs for the jet smearing technique. Table B.8 lists the preselection criteria to enrich multi-jet contributions in order to select jets used to apply the jet smearing. Tables B.9, B.10, B.11 and B.12 list the CR definitions targeting SRs A-E.

Table B.13 shows the expected yields of multi-jet production in all SRs and VRs estimated using the jet smearing technique.

Table B.8 Preselection applied for all CRs used for the estimation of multi-jet production. $\mathscr{L}$ [10^{34} cm^{-2} s^{-1}] is denoting the instantaneous luminosity

	Requirement
Number of leptons	$N_\ell = 0$
Trigger	E_T^{miss}, HLT threshold: 70 GeV (2015)
	E_T^{miss}, HLT threshold: 90 GeV (2016, $\mathscr{L} \leq 1.02$)
Missing transverse energy	$E_T^{miss} > 200$ GeV
Number of jets	$N_{jets} \geq 4$
Number of b-jets	$N_{b-jets} \geq 2$
$\left| \Delta\phi \left(\text{jet}^{0,1}, E_T^{miss} \right) \right|$	<0.4

Table B.9 Requirements for the multi-jet CRs and VRs targeting SRA and SRB in addition to the requirements presented in Table B.8

Region		CRQ	VRQ
	min $\left\lvert\Delta\phi\left(\text{jet}^{0-1}, E_T^{\text{miss}}\right)\right\rvert$	<0.1	[0.1, 0.4]
	$m^0_{\text{jet},R=1.2}$	>120 GeV	
	$m_T^{b,\text{min}}$	>100 GeV	
	b-tagged jets	≥2	
	τ-veto	Yes	
A	$m^0_{\text{jet},R=0.8}$	>60 GeV	
	E_T^{miss}	>300 GeV	
B	$m_T^{b,\text{max}}$	>200 GeV	
	$\Delta R(b,b)$	>1.2	

Table B.10 Requirements for the multi-jet CRs and VRs targeting SRC in addition to the requirements presented in Table B.8

Variable	CR	VR
min $\left\lvert\Delta\phi\left(\text{jet}^{0-1}, E_T^{\text{miss}}\right)\right\rvert$	[0.05, 0.1]	[0.1, 0.2]
$N^S_{b-\text{jet}}$	≥1	
N^S_{jet}	≥5	
$p_{T,b}^{0,S}$	>40 GeV	
m_S	–	>300 GeV
$\Delta\phi_{\text{ISR},E_T^{\text{miss}}}$	>2.00	>3.00
p_T^{ISR}	>150 GeV	>400 GeV
R_{ISR}	<0.4	
$p_T^{4,S}$	>50 GeV	
b-tagged jets	≥1	

Table B.11 Requirements for the multi-jet CRs and VRs targeting SRD-low and SRD-high in addition to the requirements presented in Table B.8

	CRQD	VRQD	VRQD-low	VRQD-high
min $\left\lvert\Delta\phi\left(\text{jet}^{0-1}, E_T^{\text{miss}}\right)\right\rvert$	< 0.1	[0.1, 0.4]		
E_T^{miss}	>250 GeV			
NJets	≥5			
b-tagged jets	≥2			
$\Delta R(b,b)$	>0.8			
τ-veto	Yes			
Jet p_T^1	–		>150 GeV	
Jet p_T^3	–		> 100 GeV	>80 GeV
Jet p_T^4	–		>60 GeV	
$m_T^{b,\text{min}}$	–		>250 GeV	>350 GeV
$m_T^{b,\text{max}}$	–		>300 GeV	>450 GeV
b-jet $p_T^0+p_T^1$	–		>300 GeV	>400 GeV

Table B.12 Requirements for the multi-jet CRs and VRs targeting SRE in addition to the requirements presented in Table B.8

Variable	CRQE	VRQE
$\min \left\| \Delta\phi \left(\mathrm{jet}^{0-1}, E_T^{\mathrm{miss}} \right) \right\|$	<0.1	[0.1, 0.4]
b-tagged jets	≥ 2	
$m_T^{b,\mathrm{min}}$	>100 GeV	
E_T^{miss}	>250 GeV	
$m_{\mathrm{jet},R=0.8}^0$	>120 GeV	
$m_{\mathrm{jet},R=0.8}^1$	>80 GeV	
H_T	>800 GeV	
$E_T^{\mathrm{miss}}/\sqrt{H_T}$	–	

Table B.13 Expected yields of the multi-jet backgrounds in all signal regions vetoing the presence of leptons estimated using the jet smearing technique

Region	Predicted multi-jet yield
SRA-TT	0.21 ± 0.10
SRA-TW	0.14 ± 0.09
SRA-T0	0.12 ± 0.07
SRB-TT	1.54 ± 0.64
SRB-TW	1.01 ± 0.88
SRB-T0	1.79 ± 1.54
SRC-1	4.56 ± 2.38
SRC-2	1.58 ± 0.77
SRC-3	0.32 ± 0.17
SRC-4	0.04 ± 0.02
SRC-5	0.00 ± 0.00
SRD-low	1.12 ± 0.37
SRD-high	0.40 ± 0.15
SRE	0.00 ± 0.00
VRTopAT0	0.14 ± 0.08
VRTopATT	0.98 ± 0.65
VRTopATW	0.12 ± 0.13
VRTopBT0	0.66 ± 0.55
VRTopBTT	2.00 ± 1.65
VRTopBTW	0.60 ± 0.77
VRTopC	2.96 ± 2.33
VRTopD	2.43 ± 0.84
VRTopE	0.00 ± 0.00
VRZAB	0.63 ± 1.65
VRZD	0.76 ± 0.57
VRZE	0.00 ± 0.00

B.8 Estimation of Systematic Uncertainties

This section contains additional information on the estimation of systematic uncertainties for the search for top squark pair production.

B.8.1 Experimental Uncertainties

Figures B.14, B.15, B.16, B.17, B.18 and B.19 show the relative experimental systematic uncertainties binned in the transverse momentum of the highest-p_T b-jet for all SRs. Thereby, the uncertainties are grouped into categories. In case more

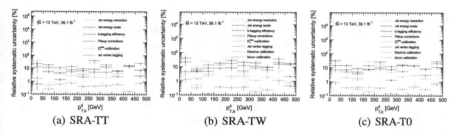

(a) SRA-TT   (b) SRA-TW   (c) SRA-T0

Fig. B.14 Relative experimental systematic uncertainties in SRA binned in the transverse momentum of the highest-p_T b-jet

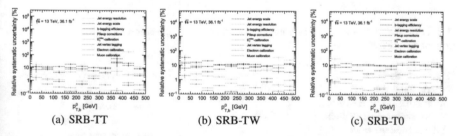

(a) SRB-TT   (b) SRB-TW   (c) SRB-T0

Fig. B.15 Relative experimental systematic uncertainties in SRB binned in the transverse momentum of the highest-p_T b-jet

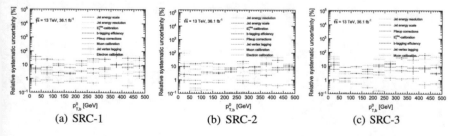

(a) SRC-1   (b) SRC-2   (c) SRC-3

Fig. B.16 Relative experimental systematic uncertainties in SRC-1, SRC-2 and SRC-3 binned in the transverse momentum of the highest-p_T b-jet

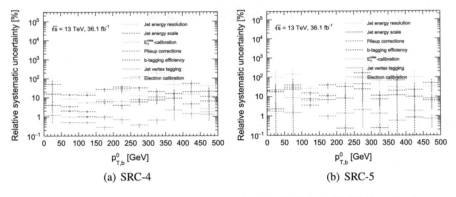

(a) SRC-4   (b) SRC-5

Fig. B.17 Relative experimental systematic uncertainties in SRC-4 and SRC-5 binned in the transverse momentum of the highest-p_T b-jet

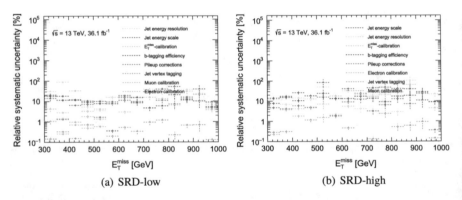

(a) SRD-low   (b) SRD-high

Fig. B.18 Relative experimental systematic uncertainties in SRD binned in the transverse momentum of the highest-p_T b-jet

Fig. B.19 Relative experimental systematic uncertainties in SRE binned in the transverse momentum of the highest-p_T b-jet

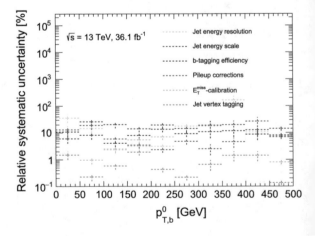

Table B.14 Overview of the simulations used for estimating the theoretical uncertainties on the MC generation of the SM processes. More details of the generator configurations can be found in [3–6]

Process	Generator	Showering	PDF set	UE tune	Order
$t\bar{t}$	SHERPA 2.2.1	SHERPA 2.2.1	NNPDF3.0NNLO	SHERPA default	NLO
$t\bar{t}$	POWHEG- BOX 2	Herwig++	CT10 (NLO)	UE5C6L1	NLO
$t\bar{t}$ (ISR/FSR x2)	POWHEG- BOX 2	PYTHIA 6.428	CT10 (NLO)	PERUGIA 2012	NNLO+NNLL
$t\bar{t}$ (ISR/FSR x0.5)	POWHEG- BOX 2	PYTHIA 6.428	CT10 (NLO)	PERUGIA 2012	NNLO+NNLL
• Wt-channel	POWHEG- BOX 2	Herwig++	CT10 (NLO)	UE5C6L1	NLO
• Wtb	MADGRAPH 2.2.3	PYTHIA 8.186	NNPDF2.3LO	A14	LO
• $WWbb$	MADGRAPH 2.2.3	PYTHIA 8.186	NNPDF2.3LO	A14	LO

Table B.15 Summary of relative theoretical uncertainties on $t\bar{t}$ production obtained on the transfer factor for all SRs and VRs

Signal region	MC generator	Parton showering	ISR/FSR
SRA-TT	–	0.90	0.53
SRA-TW	1.0897	0.20	0.11
SRA-T0	0.3530	0.30	0.32
SRB-TT	0.3704	0.39	0.33
SRB-TW	0.1839	0.20	0.33
SRB-T0	0.4281	0.21	0.22
SRC-1	0.3780	0.05	0.20
SRC-2	0.0864	0.06	0.10
SRC-3	0.1639	0.11	0.04
SRC-4	0.1379	0.11	0.10
SRC-5	–	0.20	0.05
SRD-low	0.9204	0.62	0.28
SRD-high	0.8509	1.60	0.05
SRE	2.2387	0.00	0.30
VRTopATT	0.3350	0.20	0.09
VRTopATW	0.1041	0.07	0.13
VRTopAT0	0.1558	0.08	0.04
VRTopBTT	0.2481	0.26	0.07
VRTopBTW	0.1895	0.19	0.10
VRTopBT0	0.0624	0.08	0.04
VRTopC	0.0408	0.04	0.03
VRTopD	0.1721	0.06	0.01
VRTopE	0.2962	0.27	0.04

Table B.16 Summary of relative theoretical uncertainties on single top quark production obtained on the transfer factor for all SRs and VRs

Signal region	Relative theoretical uncertainty
SRA-TT	0.172
SRA-TW	0.157
SRA-T0	0.116
SRB-TT	0.100
SRB-TW	0.100
SRB-T0	0.108
SRC-1	0.116
SRC-2	0.142
SRC-3	0.121
SRC-4	0.091
SRC-5	0.282
SRD-low	0.112
SRD-high	0.103
SRE	0.407
CRTopATT	0.097
CRTopATW	0.114
CRTopAT0	0.133
CRTopBTT	0.097
CRTopBTW	0.095
CRTopBT0	0.126
VRTopATT	0.090
VRTopATW	0.090
VRTopAT0	0.090
VRTopBTT	0.090
VRTopBTW	0.090
VRTopBT0	0.090
VRTopC	0.093
VRTopD	0.092
VRTopE	0.099
VRZAB	0.090
VRZD	0.090
VRZE	0.090
VRW	0.090

Table B.17 Quadratic sum of relative theoretical uncertainties on single Z+jets and W+jets production obtained on the transfer factor for all SRs and VRs

Signal region	Z+jets	W+jets
SRA-TT	0.0448	0.0951
SRA-TW	0.0549	0.0803
SRA-T0	0.0285	0.0607
SRB-TT	0.0406	0.0912
SRB-TW	0.0407	0.0791
SRB-T0	0.0307	0.0325
SRC-1	–	0.1141
SRC-2	–	0.1254
SRC-3	–	0.1185
SRC-4	–	0.1073
SRC-5	–	0.0949
SRD-high	0.0219	0.0820
SRD-low	0.0247	0.0882
SRE	0.0396	0.0953
VRZAB	0.0687	–
VRZD	0.0156	–
VRZE	0.0376	–
VRW	–	0.0194

than one uncertainty is part of one category, the uncertainties are added in quadrature. Throughout all SRs, the dominating experimental uncertainties are arising from the estimation of the jet energy scale, jet energy calibration, E_T^{miss} calibration and b-tagging efficiency.

B.8.2 Theoretical Uncertainties

For the estimation of theoretical uncertainties, additional events are generated for several SM and signal processes using other MC generators or generator settings. An overview on the simulations used in the estimation of the theoretical uncertainties is given in Table B.14.

Tables B.15 and B.16 list the relative theoretical uncertainties on $t\bar{t}$ and single top quark production obtained on the transfer factor for all SRs and VRs which are propagated to the simultaneous fit.

Table B.17 lists the relative theoretical uncertainties on Z+jets and W+jets quark production obtained on the transfer factor for all SRs and VRs which are propagated to the simultaneous fit.

Table B.18 Summary of the signal theory uncertainties for all SRs. The numbers are taken for signal scenarios sensitive to the respective signal regions

SUSY model	Uncertainty (%) per signal region								
	SRA-TT	SRA-TW	SRA-T0	SRB-TT	SRB-TW	SRB-T0	SRD-low	SRD-high	SRE
tN1/Mixed $\mathcal{BR} = 100\%$	10						25		15
Mixed $\mathcal{BR} = 75\%$	10			25		20	15		
Mixed $\mathcal{BR} = 50\%$	10			25		20	15		
Mixed $\mathcal{BR} = 25\%$	15		10	25		20	15		
Mixed $\mathcal{BR} = 0\%$	30	25	15	25		15	20	25	
Gtc	15			10					
Wino NLSP	20	15	20	15			15		
Well-tempered (M_{q3L})	15		20	10			15		
Well-tempered (M_{tR})	35				20		35		
Non-As. $\tilde{h}$	15			25	15	10	25		

Table B.18 lists the signal theory uncertainties for all SRs. The numbers are taken for signal scenarios sensitive to the respective signal regions.

B.9   Results of the Search for Top Squarks

This section contains additional results of the search for top squarks not presented within the main part of the thesis. Table B.19 shows the observed and predicted event yields in SRE obtained by the simultaneous fit. Figure B.20 shows a summary of the observed and predicted event yields for all SRs after the likelihood fit.

Table B.19 Fit results in SRE for an integrated luminosity of 36.1 fb^{-1}. The background normalisation parameters are obtained from the background-only fit in the CRs and are applied to the SRs. The uncertainties in the yields include statistical uncertainties and all systematic uncertainties defined in Sect. 11.5

	SRE
Observed	3
Total MC	3.64 ± 0.79
$t\bar{t}$	$0.21^{+0.39}_{-0.21}$
Z+jets	1.36 ± 0.25
W+jets	0.52 ± 0.27
Single Top	0.66 ± 0.49
$t\bar{t} + V$	0.89 ± 0.19
Diboson	–
Multi-jets	–

Fig. B.20 Observed and predicted event yields for all SRs after the likelihood fit. The stacked histograms show the SM prediction and the hatched uncertainty band around the SM prediction shows total uncertainty, which consists of the MC statistical uncertainties, detector-related systematic uncertainties, and theoretical uncertainties in the extrapolation from CR to SR [2]

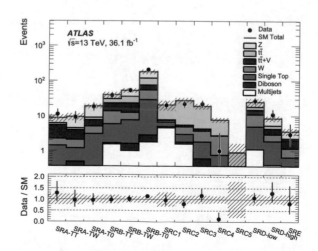

References

1. ATLAS Collaboration (2016) Measurement of $W^{\pm}$ and Z-boson production cross sections in pp collisions at $\sqrt{s} = 13$ TeV with the ATLAS detector. Phys Lett B 759:601. https://doi.org/10.1016/j.physletb.2016.06.023
2. ATLAS Collaboration (2017) Search for a scalar partner of the top quark in the jets plus missing transverse momentum final state at $\sqrt{s} = 13$ TeV with the ATLAS detector. JHEP 12:085. arXiv:1709.04183 [hep-ex]
3. ATLAS Collaboration (2016) Simulation of top-quark production for the ATLAS experiment at $\sqrt{s} = 13$ TeV. ATL-PHYS-PUB-2016-004. https://cds.cern.ch/record/2120417
4. ATLAS Collaboration (2016) Monte Carlo generators for the production of a W or Z/γ^* Boson in association with jets at ATLAS in run 2. ATL-PHYS-PUB-2016-003. https://cds.cern.ch/record/2120133
5. ATLAS Collaboration (2016) Multi-boson simulation for 13 TeV ATLAS analyses. ATL-PHYS-PUB-2016-002. https://cds.cern.ch/record/2119986
6. ATLAS Collaboration (2016) Modelling of the $t\bar{t}H$ and $t\bar{t}V(V = W, Z)$ processes for $\sqrt{s} = 13$ TeV ATLAS analyses. ATL-PHYS-PUB-2016-005. https://cds.cern.ch/record/2120826

Appendix C
Search for Dark Matter

C.1 Validation of Simulated Signal Processes Using the Fast Simulation Framework

As in the search for top squarks, the usage of simulations where the showers in the electromagnetic and hadronic calorimeters are simulated with a parametrised description instead of using GEANT 4 has to be validated against simulated events where also GEANT 4 was exploited to simulate the calorimeter showers. Figures C.1, C.2 and C.3 show a selection of simulated distributions of discriminating variables used in the search for Dark Matter normalised to an integrated luminosity of $1\,\text{fb}^{-1}$. The figures compare the simulation reconstructed with the parametrised description of the calorimeter showers (fast simulation) with the full GEANT 4 setup. All distributions are in good agreement within the statistical and experimental systematic uncertainties.

C.2 Basic Experimental Signature

Figure C.4 shows the relative distributions of the jet transverse momenta p_T for the first four jets with the highest p_T after applying a lepton veto as well as at least four jets and one b-jet.

C.3 Selection Criteria for Validation Regions

Tables C.1 and C.2 show the expected composition of the VRs for $t\bar{t}$ and Z+jets production, respectively. Besides the event yield observed in data, the purities of the respective contribution of interest as well as the normalisation factors computed with Eq. (10.1) are shown.

© Springer Nature Switzerland AG 2019
N. M. Köhler, *Searches for the Supersymmetric Partner of the Top Quark, Dark Matter and Dark Energy at the ATLAS Experiment*, Springer Theses, https://doi.org/10.1007/978-3-030-25988-4

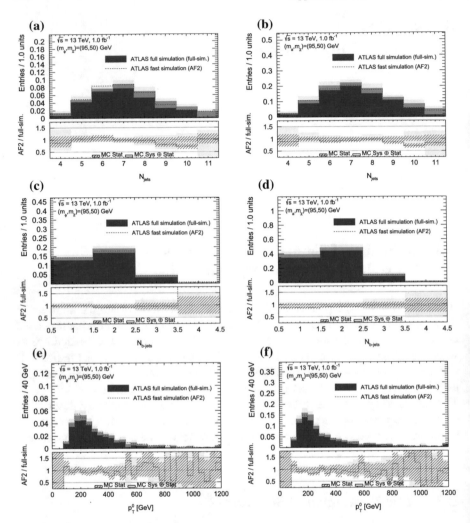

Fig. C.1 Comparison between the full ATLAS detector simulation based on GEANT 4 (red) and the fast simulation framework (AF2) for WIMP pair production with $m_\chi = 50$ GeV and scalar (left) or pseudoscalar (right) mediator with $m_{\phi/a} = 95$ GeV. The hashed (yellow) bands are indicating the statistical (experimental systematic) uncertainties

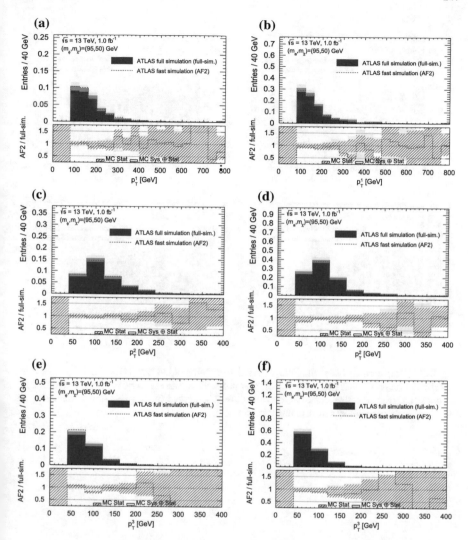

Fig. C.2 Comparison between the full ATLAS detector simulation based on GEANT 4 (red) and the fast simulation framework (AF2) for WIMP pair production with $m_\chi = 50$ GeV and scalar (left) or pseudoscalar (right) mediator with $m_{\phi/a} = 95$ GeV. The hashed (yellow) bands are indicating the statistical (experimental systematic) uncertainties

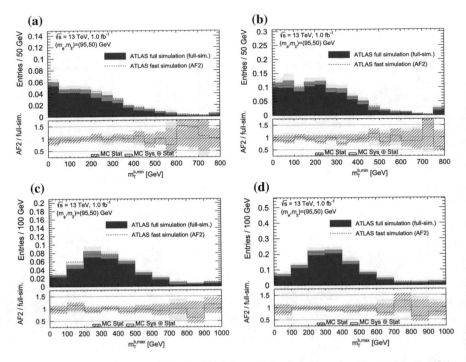

Fig. C.3 Comparison between the full ATLAS detector simulation based on GEANT 4 (red) and the fast simulation framework (AF2) for WIMP pair production with $m_\chi = 50$ GeV and scalar (left) or pseudoscalar (right) mediator with $m_{\phi/a} = 95$ GeV. The hashed (yellow) bands are indicating the statistical (experimental systematic) uncertainties

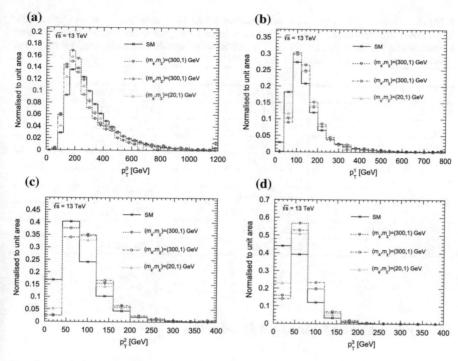

Fig. C.4 Relative distributions of the jet transverse momenta p_T for the first four jets with the highest p_T after applying a lepton veto as well as at least four jets and one b-jet for the same selection of signal processes shown in Fig. 12.3

Table C.1 Composition of SM processes in $t\bar{t}$ validation regions for $36.1\,\mathrm{fb}^{-1}$ of pp collision data. None of the simulated SM predictions is normalised using normalisation factors derived from a simultaneous fit (detailed in Sect. 11.6). The purities of the simulated $t\bar{t}$ contributions as well as the normalisation factors computed with Eq. (10.1) are also shown for all regions

	VRTt1	VRTt2
$t\bar{t}$	79.64 ± 1.83	119.65 ± 2.42
$t\bar{t} + V$	2.72 ± 0.33	6.54 ± 0.43
Z+jets	3.60 ± 0.36	9.17 ± 0.77
W+jets	2.62 ± 0.67	4.78 ± 1.11
Single Top	3.24 ± 0.51	3.80 ± 0.42
Diboson	0.49 ± 0.17	1.69 ± 0.46
Total MC	92.31 ± 2.08	145.62 ± 2.87
Data	113.00 ± 10.63	165.00 ± 12.85
Purity of $t\bar{t}$ [%]	86.28 ± 2.78	82.17 ± 2.32
Normalisation of $t\bar{t}$	1.26 ± 0.14	1.16 ± 0.12

Table C.2 Composition of SM processes in Z+jets validation regions for 36.1 fb^{-1} of pp collision data. None of the simulated SM predictions is normalised using normalisation factors derived from a simultaneous fit (detailed in Sect. 11.6). The purities of the simulated Z+jets contributions as well as the normalisation factors computed with Eq. (10.1) are also shown for all regions

	VRZt1	VRZt2
$t\bar{t}$	161.94 ± 3.95	85.75 ± 3.46
$t\bar{t} + V$	14.27 ± 0.65	7.65 ± 0.49
Z+jets	148.90 ± 3.55	123.47 ± 3.40
W+jets	103.16 ± 4.35	74.58 ± 3.84
Single Top	66.56 ± 2.00	34.41 ± 1.49
Diboson	20.34 ± 1.83	16.87 ± 1.59
Total MC	515.17 ± 7.41	342.73 ± 6.57
Data	748.00 ± 27.35	409.00 ± 20.22
Purity of Z+jets [%]	28.90 ± 0.80	36.03 ± 1.21
SF of Z+jets	2.56 ± 0.21	1.54 ± 0.19

C.4 Estimation of Multi-jet Production

Table C.3 details the selections applied in the common CR and VR for multi-jet production.

Table C.3 Selection criteria for the CR and VR for multi-jet production targeting the DM SRs

Variable	CRQCD	VRQCD
$\left\lvert \Delta\phi \left(\mathrm{jet}^{0-3}, E_\mathrm{T}^\mathrm{miss} \right) \right\rvert$	<0.1	$[0.1, 0.3]$
$E_\mathrm{T}^\mathrm{miss}/\sqrt{H_\mathrm{T}}$	–	$<10 \sqrt{\mathrm{GeV}}$
$E_\mathrm{T}^\mathrm{miss}$	$>250\,\mathrm{GeV}$	
N_lep	0	
Jet transverse momenta	$\geq 4, \ p_\mathrm{T} > 80, 80, 40, 40\,\mathrm{GeV}$	
b-tagged jets	≥ 2	
$E_\mathrm{T}^\mathrm{miss,track}$	$>30\,\mathrm{GeV}$	
$\left\lvert \Delta\phi \left(E_\mathrm{T}^\mathrm{miss}, E_\mathrm{T}^\mathrm{miss,track} \right) \right\rvert$	$<\pi/3$	
τ veto	No	Yes
$m_{\mathrm{jet},R=0.8}^{0}$	$>80\,\mathrm{GeV}$	$>40\,\mathrm{GeV}$
$m_{\mathrm{jet},R=0.8}^{1}$	$>80\,\mathrm{GeV}$	$>40\,\mathrm{GeV}$
$m_\mathrm{T}^{b,\mathrm{max}}$	$>100\,\mathrm{GeV}$	
$\Delta R\,(b, b)$	–	>1.5

Appendix D
Search for Dark Energy

D.1 Basic Experimental Signature

Figure D.1 shows the relative distributions of the jet transverse momenta p_T for the first four jets with the highest p_T after applying a lepton veto as well as at least four jets and one b-jet.

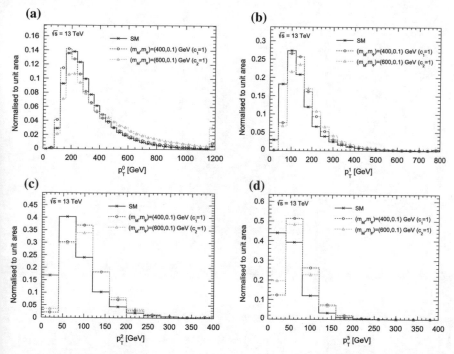

Fig. D.1 Relative distributions of the jet transverse momenta p_T for the first four jets with the highest p_T after applying a lepton veto as well as at least four jets and one b-jet for the same selection of signal processes shown in Fig. 13.2

© Springer Nature Switzerland AG 2019
N. M. Köhler, *Searches for the Supersymmetric Partner of the Top
Quark, Dark Matter and Dark Energy at the ATLAS Experiment*,
Springer Theses, https://doi.org/10.1007/978-3-030-25988-4

Appendix E
Multivariate Techniques in the Search for Top Squarks

E.1 Boosted Decision Trees

High Top Squark Masses

Figures E.1, E.2, E.3 and E.4 show the distributions of kinematic variables used in the BDT training for SRs A, B and D after the selection requirements summarised in Table 14.1. For the bins with significant statistics no difference in the shape of the data with respect to the SM prediction is observed. The differences in the normalisation arise since no normalisation factors estimated in CRs are applied. However, since the multivariate techniques are only considering the shape of the discriminating variables, the differences are expected to have a minor effect. Figures E.5 and E.6 show a comparison between the expected exclusion limits at 95% CL obtained by the nominal maximum likelihood fit exploiting all CRs and systematic uncertainties and a simplified maximum likelihood fit assuming an overall systematic uncertainty of 30% and no dedicated CRs for SRA-TT and a statistical combination of SRB-TT, SRB-TW and SRB-T0, respectively. The simplified maximum likelihood fit assuming an overall systematic uncertainty of 30% and no dedicated CRs can be used as a conservative measure for determining expected exclusion limits in the optimisation studies exploiting multivariate techniques. Figures E.7 and E.8 show the correlation matrices for signal and background events for the BDTs targeting SRA and SRB, respectively.

Compressed Scenarios

Figures E.9 and E.10 show the distributions of kinematic variables used in the BDT training for SRC after the selection requirements summarised in Table 14.3. For the bins with significant statistics no difference in the shape of the data with respect to the SM prediction is observed. The differences in the normalisation arise since no

© Springer Nature Switzerland AG 2019
N. M. Köhler, *Searches for the Supersymmetric Partner of the Top
Quark, Dark Matter and Dark Energy at the ATLAS Experiment*,
Springer Theses, https://doi.org/10.1007/978-3-030-25988-4

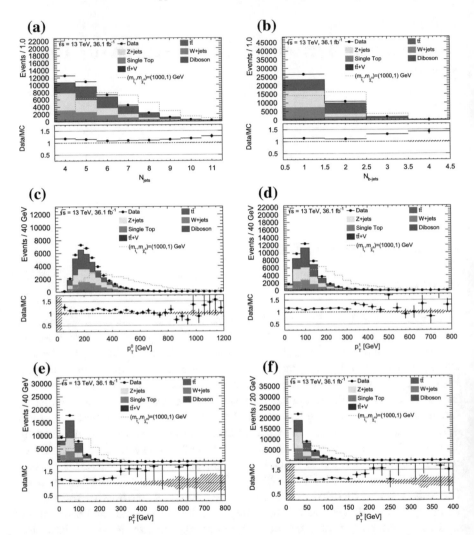

Fig. E.1 Distributions of kinematic variables used in the BDT training for SRs A, B and D after the selection requirements summarised in Table 14.1. The stacked histograms show the SM expectation whereas the dashed line shows the simulated distribution of top squark pair production with $m_{\tilde{t}_1} = 1000$ GeV and $m_{\tilde{\chi}_1^0} = 1$ GeV normalised to the integral of the SM expectation

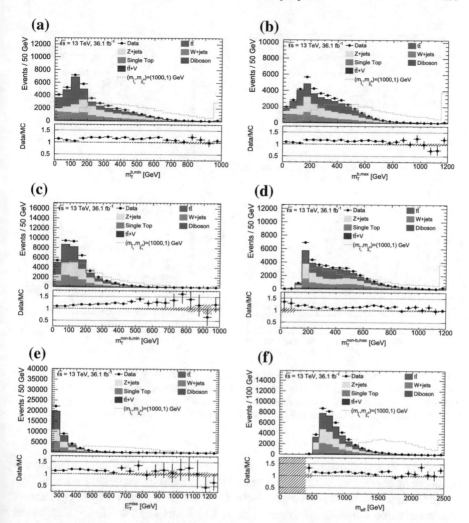

Fig. E.2 Distributions of kinematic variables used in the BDT training for SRs A, B and D after the selection requirements summarised in Table 14.1. The stacked histograms show the SM expectation whereas the dashed line shows the simulated distribution of top squark pair production with $m_{\tilde{t}_1} = 1000\,\text{GeV}$ and $m_{\tilde{\chi}_1^0} = 1\,\text{GeV}$ normalised to the integral of the SM expectation

normalisation factors estimated in CRs are applied. However, since the multivariate techniques are only considering the shape of the discriminating variables, the differences are expected to have a minor effect. Figure E.11 shows a comparison between the expected exclusion limits at 95% CL obtained by the nominal maximum likelihood fit exploiting all CRs and systematic uncertainties and a simplified maximum likelihood fit assuming an overall systematic uncertainty of 20% and no dedicated CRs for a statistical combination of SRC1-5. The simplified maximum

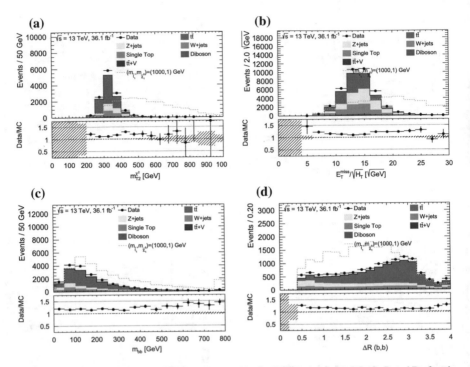

Fig. E.3 Distributions of kinematic variables used in the BDT training for SRs A, B and D after the selection requirements summarised in Table 14.1. The stacked histograms show the SM expectation whereas the dashed line shows the simulated distribution of top squark pair production with $m_{\tilde{t}_1} = 1000\,\text{GeV}$ and $m_{\tilde{\chi}_1^0} = 1\,\text{GeV}$ normalised to the integral of the SM expectation

likelihood fit assuming an overall systematic uncertainty of 30% and no dedicated CRs can be used as a conservative measure for determining expected exclusion limits in the optimisation studies exploiting multivariate techniques. Figure E.12 shows the comparison between the expected exclusion limits at 95% CL for a SR based on the preselection summarised in Table 14.3 and trained on all signal simulations with $173\,\text{GeV} \leq m_{\tilde{t}_1} - m_{\tilde{\chi}_1^0} \leq 188\,\text{GeV}$ (a) as well as only using simulations with $m_{\tilde{t}_1} \leq 400\,\text{GeV}$. The usage of signal scenarios with low top squark masses only does not result in an improvement of the expected exclusion limits towards lower masses.

The correlation matrices for signal and background events (cf. Fig. 14.9) show a strong correlation between the transverse momenta of the jets in the sparticle system with m_S and p_T^{ISR} as well as a strong anti-correlation with respect to R_{ISR} for both background and signal events. Figure E.13 shows the simulated distributions of those variables for both the SM prediction and the signal simulations used in the training of the BDT. The correlations are indeed identical for background and signal events which results in the BDT having less discriminating power and thus not providing improved expected exclusion limits.

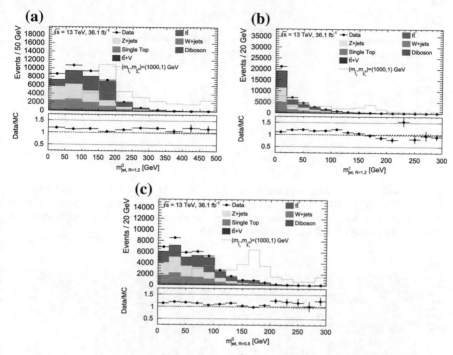

Fig. E.4 Distributions of kinematic variables used in the BDT training for SRs A, B and D after the selection requirements summarised in Table 14.1. The stacked histograms show the SM expectation whereas the dashed line shows the simulated distribution of top squark pair production with $m_{\tilde{t}_1} = 1000\,\text{GeV}$ and $m_{\tilde{\chi}_1^0} = 1\,\text{GeV}$ normalised to the integral of the SM expectation

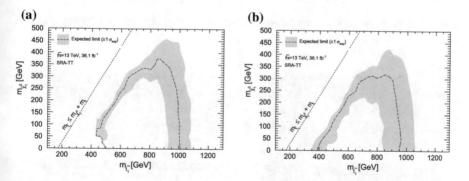

Fig. E.5 Comparison between the expected exclusion limits at 95% CL for SRA-TT obtained by the nominal maximum likelihood fit exploiting all CRs and systematic uncertainties (**a**) and a simplified maximum likelihood fit assuming an overall systematic uncertainty of 30% and no dedicated CRs (**b**)

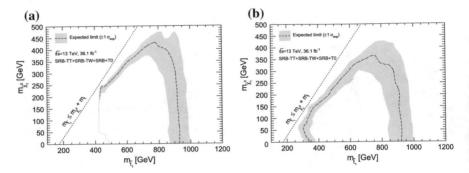

Fig. E.6 Comparison between the expected exclusion limits at 95% CL for the combination of SRB-TT, SRB-TW and SRB-T0 obtained by the nominal maximum likelihood fit exploiting all CRs and systematic uncertainties (**a**) and a simplified maximum likelihood fit assuming an overall systematic uncertainty of 30% and no dedicated CRs (**b**)

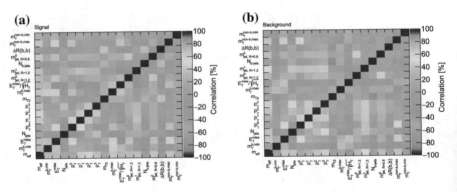

Fig. E.7 Correlation matrices for signal (**a**) and background (**b**) events for the BDT targeting SRA

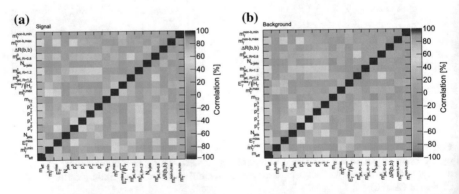

Fig. E.8 Correlation matrices for signal (**a**) and background (**b**) events for the BDT targeting SRB

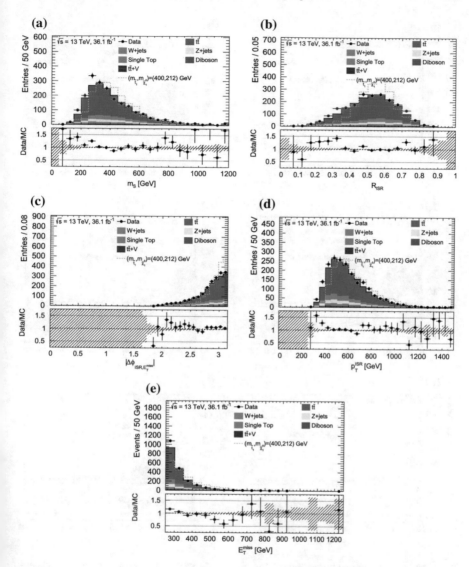

Fig. E.9 Distributions of kinematic variables used in the BDT training for SRC after the selection requirements summarised in Table 14.3. The stacked histograms show the SM expectation whereas the dashed line shows the simulated distribution of top squark pair production with $m_{\tilde{t}_1} = 400\,\text{GeV}$ and $m_{\tilde{\chi}_1^0} = 212\,\text{GeV}$ normalised to the integral of the SM expectation

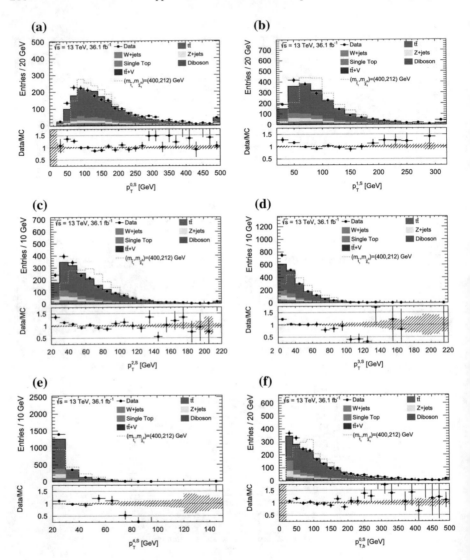

Fig. E.10 Distributions of kinematic variables used in the BDT training for SRC after the selection requirements summarised in Table 14.3. The stacked histograms show the SM expectation whereas the dashed line shows the simulated distribution of top squark pair production with $m_{\tilde{t}_1} = 400\,\text{GeV}$ and $m_{\tilde{\chi}_1^0} = 212\,\text{GeV}$ normalised to the integral of the SM expectation

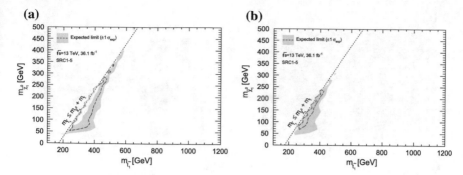

Fig. E.11 Comparison between the expected exclusion limits at 95% CL for the combination of SRC1-5 obtained by the nominal maximum likelihood fit exploiting all CRs and systematic uncertainties (**a**) and a simplified maximum likelihood fit assuming an overall systematic uncertainty of 20% and no dedicated CRs (**b**)

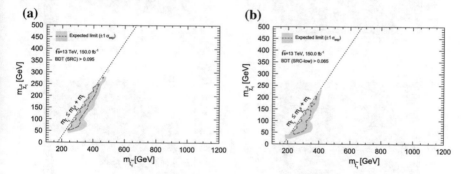

Fig. E.12 Comparison between the expected exclusion limits at 95% CL for a SR based on the preselection summarised in Table 14.3 and trained on all signal simulations with $173\,\text{GeV} \leq m_{\tilde{t}_1} - m_{\tilde{\chi}_1^0} \leq 188\,\text{GeV}$ (**a**) as well as only using simulations with $m_{\tilde{t}_1} \leq 400\,\text{GeV}$ (**b**). An overall systematic uncertainty of 20% is assumed and no dedicated CRs are defined

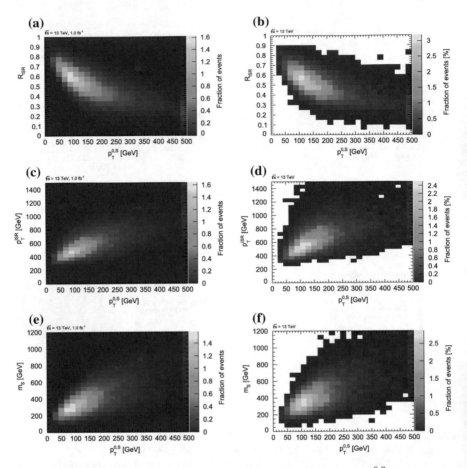

Fig. E.13 Simulated distributions of the leading jet of the sparticle system, $p_T^{0,S}$, drawn against R_{ISR}, p_T^{ISR} and m_S for the SM expectation (left) and the signal simulations (right) used in the BDT training

E.2 Artificial Neural Networks

Figure E.14 shows the output of the Kolmogorov–Smirnov test depending on the choice of the maximum number of iterations for the training and the number of neurons per layer for background and signal events. For the choice of the configuration parameters of 200 iterations and two hidden layers containing 150 and 10 neurons overtraining is excluded by a level > 95%. Figure E.15 shows the correlation matrices for signal and background events for the NN targeting SRA.

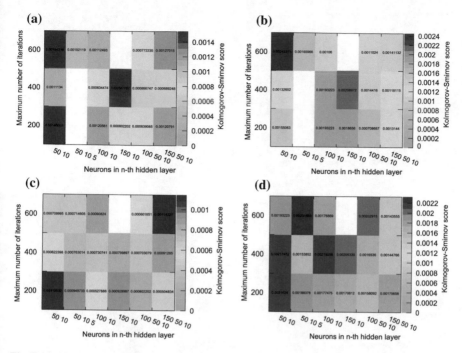

Fig. E.14 Kolmogorov–Smirnov test depending on the choice of the maximum number of iterations for the training and the number of neurons per layer for background (left) and signal (right) events. The results of the nominal training and testing phase are shown in the top row, the bottom row shows the result, if the events used for training and testing, respectively are swapped. Since the number of background (signal) events used in the training and testing, respectively, was 433,762 (165,094), for the choice of the configuration parameters of 200 iterations and two hidden layers containing 150 and 10 neurons overtraining is excluded by a level $> 95\%$

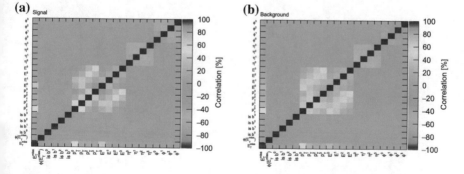

Fig. E.15 Correlation matrices for signal (**a**) and background (**b**) events for the NN targeting SRA